Bat Conse

Bat Surveys for Professional Ecologists

Good Practice Guidelines (3rd edition)

Foreword

Bat Surveys for Professional Ecologists: Good Practice Guidelines (3rd edn) builds on previous editions using feedback from general comments to the Bat Conservation Trust and a public consultation, following the publication of the second edition in 2012. Representatives of developers, ecological consultancies, local authorities, licensing authorities, academic institutions and voluntary bat workers have provided comments, which have been collated and considered in the writing of this edition. The comments highlighted areas of the existing guidelines that needed clarification; new subject areas that should be added; and necessary updates following changes in legislation, licensing, policy or the publication of new research. A Technical Review Board, consisting of a wide range of stakeholders, has reviewed this edition. This document is available to purchase as a hard copy, or can be downloaded from www.bats.org.uk. This edition will be subject to review after two years; any comments should be sent to surveyguidelines@bats.org.uk.

Julia Hanmer
Joint Chief Executive

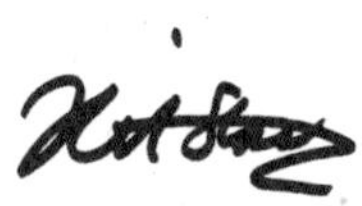

Kit Stoner
Joint Chief Executive

Acknowledgements

This publication has benefitted greatly from the input of a large number of people. In particular, the Bat Conservation Trust would like to extend its thanks to the new contributing authors, Technical Review Board and technical reviewers/contributors, whose names are all listed on the next page. All of these people gave significant amounts of their time free of charge to write new material and review draft versions of the document.

BCT would like to thank all those who responded to the consultation held in 2013 and those who have provided comments to BCT since the publication of the second edition of these guidelines. All comments and suggestions were considered and many were incorporated into this version. Space precludes us from listing everyone here but their input was invaluable.

Several members of BCT staff have contributed a great deal to this document in a variety of ways, including Kate Barlow, Katherine Boughey, Pete Charleston, Jo Ferguson, Lisa Hundt, Helen Miller and Carol Williams, who provided a sounding board, wrote or discussed sections of the text, and reviewed draft versions of the document.

Many thanks to our sponsors for providing funding towards the printing costs, thereby enabling proceeds from the sale of this document to go towards bat conservation.

Many others have helped with the production of this document and it has not been possible to list everyone by name. We would like to thank you all for your time and expertise.

Editor
Jan Collins (BCT)

Contributing authors

Jan Collins (BCT): all chapters/sections except those listed below.

Pete Charleston (BCT): *Section 1.2.1, Legislative context.*

Ian Davidson-Watts (Davidson-Watts Ecology Ltd): *Chapter 9, Advanced licence bat survey techniques and Section 10.4, Analysis of bat radiotelemetry survey data.*

Steve Markham (Marquis & Lord): *Section 10.3, Analysis of bat activity survey data; Appendix 7, Introduction to data analysis; Appendix 8, Worked examples of statistical analysis, plus various shorter sections on sampling and analysis.*

Lisa Kerslake (Swift Ecology Ltd): *Appendix 4, Protocol for bat dropping collection for DNA analysis.*

Technical Review Board

The role of this Technical Review Board was to review a draft of Bat Surveys for Professional Ecologists: Good Practice Guidelines (3rd edn) produced by the authors listed above. The Technical Review Board provided comments on the draft and contributed further through verbal and written discussions on key areas. All comments and discussions were taken into account in producing the final version of these guidelines but, where consensus could not be reached, BCT took the final editorial decision. All authors of Bat Surveys – Good Practice Guidelines, 2nd edn (Hundt, 2012) were included in this Technical Review Board. Members of the Technical Review Board and their affiliations are listed below.

Sally Blyth (Scottish Natural Heritage, SNH)
Rebecca Collins (Chartered Institute for Ecology and Environmental Management, CIEEM)
Richard Crompton (Wildwood Ecology)
Matt Dodds (Herts and Middlesex Wildlife Trust)
Jo Ferguson (BCT)
Richard Graves (Richard Graves Associates)
Sean Hanna (Natural England, NE)
Matthew Hobbs (BSG Ecology)
Michelle Henley (SNH)
Lisa Hundt (BCT)
Lisa Kerslake (CIEEM)
Pól Mc Cana (Northern Ireland Environment Agency)
Louise Mapstone (CIEEM)
Jean Matthews (Natural Resources Wales, NRW)
Steve Markham (Marquis & Lord)
John McKinnell (SNH)
Helen Miller (BCT)
Mike Oxford (Association of Local Government Ecologists)
Paola Reason (Arcadis)
Peter Shepherd (BSG Ecology)
Sandie Sowler (Consultant ecologist, trainer and advisor)
Claire Storey (NE)
Kat Walsh (NE)
Carol Williams (BCT)
Claire Wilmer (CIEEM)

Specialist reviewers/contributors

Specialist reviewers/contributors were brought in to review and contribute to specific technical sections of these guidelines. They provided comments (including suggested alternative text) on drafts that were taken into account in producing the final version of these guidelines but, where consensus could not be reached, BCT took a final editorial decision. Specialist reviewers/contributors, their affiliations and the sections they reviewed/contributed to are listed below.

John Altringham (Leeds University): *Section 7.3, Swarming surveys.*

Ian Davidson-Watts (Davidson-Watts Ecology Ltd): *Section 7.3, Swarming surveys.*

Daniel Whitby (AEWC): *Chapter 8, Advanced licence bat survey techniques; Section 7.3, Swarming surveys.*

Contents

List of figures

List of tables

List of abbreviations used in text

ASSIs	Areas of Special Scientific Interest (Northern Ireland designation)
BCA	British Caving Association
BCT	Bat Conservation Trust
BSI	British Standards Institution
BS42020	British Standard 42020:2013 Biodiversity. Code of practice for planning and development
CCW	Countryside Council for Wales (now Natural Resources Wales)
CIEEM	Chartered Institute for Ecology and Environmental Management (formerly the Institute for Ecology and Environmental Management)
CITB	Construction Industries Training Board
CSCS	Construction Site Certification Scheme
CSZ	Core Sustenance Zone
DCLG	Department for Communities and Local Government
DOE	Department of the Environment (in Northern Ireland)
EC Habitats Directive	Council Directive 92/43/EEC 1992 on the conservation of natural habitats and of wild fauna and flora
EcIA	Ecological Impact Assessment
EIA	Environmental Impact Assessment
EN	English Nature (now Natural England)
EPS	European Protected Species
FC	Forestry Commission
FCS	Favourable Conservation Status
HRA	Habitats Regulations Assessment
HSE	Health and Safety Executive
IEEM	Institute for Ecology and Environmental Management (now the Chartered Institute for Ecology and Environmental Managment)
JNCC	Joint Nature Conservation Committee
LBG	local bat group
LPA	Local Planning Authority
LRC	Local Records Centre
LWT	local Wildlife Trust
MAGIC	Multi Agency Geographic Information for the Countryside
MEWP	mobile elevating work platform
NBN	National Biodiversity Network
NE	Natural England (formerly English Nature)
NERC Act	Natural Environment and Rural Communities Act, 2006
NFBR	National Forum for Biological Recording
NGO	non-governmental organisation
NNR	National Nature Reserve
NPPG	National Planning Policy Guidance
NRW	Natural Resources Wales (formerly the Countryside Council for Wales, Environment Agency Wales and Forestry Commission Wales)
PIT	passive inductor transponder
PPE	personal protective equipment
PRF	Potential Roost Feature
RIBA	Royal Institute of British Architects
SAC	Special Area of Conservation
SE	Scottish Executive
SNCO	Statutory Nature Conservation Organisation
SNH	Scottish Natural Heritage
SSSI	Site of Special Scientific Interest
ZoI	zone of influence

Chapter 1

Background

1.1 Introduction

1.1.1 Aim of the guidelines

This publication aims to provide good practice guidelines in relation to designing and undertaking bat surveys; analysing the data collected during those surveys; and writing survey reports. The guidelines relate to professional bat surveys carried out to assess how proposed activities may impact bats. The guidelines aim to raise standards and increase the consistency of this type of work and ultimately lead to a greater understanding of bats and improvements in their protection and conservation.

1.1.2 Intended audience

These guidelines are intended primarily for professional ecologists carrying out bat surveys and writing reports in relation to proposed activities that could impact bats. They may also be useful to:

- developers commissioning bat surveys and reports from ecologists in relation to development; and
- planners, ecologists and policy-makers working for local authorities, licensing authorities and non-governmental organisations (NGOs), who are responsible for reviewing and assessing the implications of professional bat surveys.

1.1.3 What the guidelines do not aim to do

The guidelines do not aim to either override or replace knowledge and experience. It is accepted that departures from the guidelines (e.g. either decreasing or increasing the number of surveys carried out or using alternative methods) are often appropriate. However, in this scenario an ecologist should provide documentary evidence of (a) their expertise in making this judgement and (b) the ecological rationale behind the judgement.

Equally, it would be inappropriate for someone with no knowledge or experience to read these guidelines and expect to be able to design, carry out, interpret the results of and report on professional surveys as a result, simply following the guidelines without the ability to apply any professional judgement. Training and experience is necessary to carry out all of the surveys described in these guidelines and interpret the survey results appropriately (see Section 2.5.1).

British Standard 42020 Biodiversity. Code of practice for planning and development (British Standards Institution (BSI), 2013, hereafter referred to as BS42020) is relevant to the planning process, other consented development and proposals involving the management and use of land. This states that:

- 'any individual dealing with ecological issues at any stage of the planning application process should be able to demonstrate that they have sufficient technical competence and experience to carry out the particular tasks and activities for which they are responsible in the role that they are performing' (BS42020; Clause 4.3.2);
- 'an explanation, with evidence, of the assessment and decision-making process and the reasons for a particular course of action or piece of advice should be clearly documented and made available where required and/or necessary' (BS42020; Clause 4.4.3); and
- 'it is especially important to provide evidence of how professional judgement has been applied where ecological work does not follow, in full or in part, the recommendations set out in national good practice guidelines' (BS42020; Note for Clause 4.4.3).

The guidelines should be interpreted and adapted on a case-by-case basis according to site-specific factors and the professional judgement of an experienced ecologist. Where examples are used in the guidelines, they are descriptive rather than prescriptive.

The guidelines do not aim to provide information on carrying out Ecological Impact Assessments (EcIAs). However, the survey work undertaken should be designed to answer questions that the impact assessment process will generate. Frequent reference is therefore made to the potential impacts of a project and associated relevant questions. *Guidelines for Ecological Impact Assessment in the UK and Ireland: Terrestrial, Freshwater and Coastal* (Chartered Institute of Ecology and Environmental Management (CIEEM), 2016) provides more information in this respect.

The guidelines do not aim to provide information on designing strategies to mitigate for impacts on bats. The *Bat Mitigation Guidelines* (Mitchell-Jones, 2004) or resources such as the Bat Conservation Trust's (BCT) Roost website can be used for this purpose.[1]

Although the survey techniques described are also often used in bat conservation or research, the guidelines have not been written for these purposes and should not be used to design such surveys. Surveys for bat conservation purposes are described in the *Bat Worker's Manual* (Mitchell-Jones and McLeish, 2004) and surveys for research purposes should be bespoke, designed according to the specific questions the research is intended to answer.

[1] http://roost.bats.org.uk/

Chapter 9, on advanced licence bat survey techniques, does not cover the use of bat rings/bands used for long-term monitoring programmes or other techniques usually associated with research such as light-tagging or passive inductor transponder (PIT) tags as these are not generally considered appropriate for surveys associated with developments. For further information on these methods, refer to Kunz and Parsons (2009).

In these guidelines, a survey is defined as a sampling activity in which a wide range of variables are measured to describe a site or an area. Surveying is distinct from monitoring, which involves repeated sampling, either year-on-year or periodically, usually to quantify changes over time or to assess whether a particular objective or standard has been attained. These guidelines do not include surveys carried out for monitoring purposes. Some information about monitoring the success of mitigation measures is provided in the *Bat Mitigation Guidelines* (Mitchell-Jones, 2004).

Please note that due to the delay in publication of the National Bats and Wind Turbines Project report, a specific chapter on wind farms is not included in this edition. Chapter 10 of the second edition of these guidelines (Hundt, 2012) will stand until new guidelines are available for this project type.

Finally, this edition of the guidelines does not include specific advice in relation to road and rail schemes, although the principles of survey design and execution do apply. Berthinussen and Altringham (2015) provide information on pre- and post-construction surveys of linear infrastructure schemes, designed specifically to assess the effectiveness of mitigation for bats crossing them.

1.2 Context for bat survey work

1.2.1 Legislative context

General, rather than comprehensive text on the legislation relating to bats and bat surveys is provided here. When dealing with individual cases, readers should consult the full texts of the relevant legislation and obtain legal advice if necessary. They should also check regularly for changes to legislation, guidance and case law. A summary of the relevant nature conservation legislation (correct at time of publication) is given in Table 1.1.

Table 1.1 Summary of the main legislation pertaining to the protection of bats in the UK.

	Habitats Regulations (transposing the EC Habitats Directive)	Other nature conservation legislation
England and Wales	Conservation of Habitats and Species Regulations 2010 (as amended)	Wildlife and Countryside Act 1981 (as amended)
Northern Ireland	Conservation (Natural Habitats, etc.) Regulations (Northern Ireland) 1995 (as amended)	Environment (Northern Ireland) Order 2002
Scotland	Conservation (Natural Habitats, &c.) Regulations 1994 (as amended)	

The EC Habitats Directive and respective domestic legislation

Annex II of the Council Directive 92/43/EEC 1992 on the conservation of natural habitats and of wild fauna and flora (EC Habitats Directive) lists animal and plant species of Community interest, the conservation of which requires the designation of Special Areas of Conservation (SACs); Annex IV lists animal and plant species of Community interest in need of strict protection. All bat species are listed in Annex IV; some are listed in Annex II.

In the UK,[2] the EC Habitats Directive has been transposed into national laws by means of the Conservation of Habitats and Species Regulations 2010 (as amended) (England and Wales),[3] the Conservation (Natural Habitats, &c.) Regulations 1994 (as amended) (Scotland)[4] and the Conservation (Natural Habitats, etc.) Regulations (Northern Ireland) 1995[5] (as amended).

Commonly the regulations are referred to as the Habitats Regulations.[6] They will now be referred to as such.

Legal framework

Although the precise wording of the legal protection afforded to bats differs between countries in the UK it all falls within a common framework making unlawful specific actions against bats but with differing emphasis on the state of mind needed to evidence offences. The legislation does not, in the main, mention bats except in annexes and schedules. The Habitats Regulations refer to specimens of European Protected Species (EPS). All species of bats found in the wild in the UK are EPS.

Kill, injure, capture/take bats

It is unlawful to kill, capture, injure or take a wild bat anywhere in the UK. In England, Wales and Northern Ireland the offence requires a deliberate action; in Scotland it requires a deliberate or reckless action. All offences of this nature are identified within the Habitats Regulations.

Disturbing bats

It is unlawful to disturb bats anywhere (roosts, flight lines or foraging areas) in particular if the level of disturbance can be

[2] The EC Habitats Directive does not apply to the Isle of Man and the Channel Islands, which are part of the British Isles but not part of the UK.
[3] http://www.legislation.gov.uk/uksi/2010/490/contents/made
[4] http://www.legislation.gov.uk/uksi/1994/2716/made
[5] http://www.legislation.gov.uk/nisr/1995/380/contents/made
[6] In Scotland and Northern Ireland the Habitats Regulations have been amended on a number of occasions, most particularly in 2007.

shown to impair their ability to survive, to breed or reproduce, to rear or nurture their young, to hibernate or migrate or to affect significantly local distribution or abundance. In England, Wales and Northern Ireland the offence requires a deliberate action. In Scotland the offence requires a deliberate or reckless action. All offences of this nature are identified within the Habitats Regulations.

In Scotland it is also an offence in the regulations to deliberately or recklessly disturb a bat whilst it is occupying a place of shelter or protection. This offence does not require the level of disturbance to be significant.

In England and Wales it is also an offence under the Wildlife and Countryside Act[7] to intentionally or recklessly disturb a bat, whilst it is occupying a place of shelter or protection. A householder who disturbs a bat in its place of shelter or protection does not commit an offence if they first seek the advice of Natural England (NE) or Natural Resources Wales (NRW) and allow time for such advice to be provided. If the bat is in the living area of a dwelling house it is not an offence in any circumstance to disturb it. This provision does not apply to Scotland or to Northern Ireland.

Harassing bats
In Scotland only it is an offence to deliberately or recklessly harass a bat or a group of bats.

Damage or destruction of roosts
Throughout the UK it is illegal to damage or destroy a place used by a bat for breeding or resting. All offences of this nature are identified within the Habitats Regulations. This offence is unique in that it can be committed accidently. No element of intentional, reckless or deliberate action needs to be evidenced.

Obstructing access to a breeding site or resting place
In Scotland it is an offence under the regulations to deliberately or recklessly obstruct access to a breeding site or resting place of a bat or to otherwise deny a bat the use of such a place. In Northern Ireland it is an offence under the regulations to deliberately obstruct access to a breeding site or resting place used by a bat.

In England and Wales it is an offence under the Wildlife and Countryside Act to intentionally or recklessly obstruct access to any place used by a bat for shelter or protection. A householder will not commit an offence if he obstructs access to a bat roost in a dwelling house providing they first seek the advice of NE or NRW and allow them time to provide such advice. This defence does not apply in Scotland or to Northern Ireland.

Possession and sale of bats
Under the Habitats Regulations it is an offence to be in possession or control of a bat alive or dead (or any part of a bat or anything derived from a bat, although bat droppings are generally considered to be acceptable), or to transport a bat, to sell or exchange a bat or to offer to sell or exchange a bat taken from the wild.

It is an offence under the Wildlife and Countryside Act in England and Wales to offer or expose for sale any bat of a species listed in Schedule 5 and taken from the wild or to possess any bat or anything derived from a bat for the purposes of sale. To publish or cause to be published any advertisement offering to buy or sell a bat.

Illegal methods for taking or killing bats
The Habitats Regulations in all parts of the UK contain provisions prohibiting certain methods of taking or killing bats even when the activity itself has been licensed. The Wildlife and Countryside Act contains similar provisions that still apply in England and Wales.

Offences relating to licensing
Actions, which would otherwise be illegal, can be made lawful if licensed by the appropriate Statutory Nature Conservation Organisation (SNCO).[8] It is an offence anywhere in the UK to make a false statement in order to obtain a bat licence or to fail to comply with the conditions of a bat licence.

Attempts and possession of items to be used to commit offences
It is an offence in all parts of the UK to attempt to commit any criminal offence or to possess items to be used to commit offences identified in any of the legislation referred to above. Legislation throughout the UK is such that it may not be only those who are directly responsible for offences that are liable. In Scotland those who cause or permit offences are guilty as are those who aid or abet offences elsewhere.

Defences
It is not illegal anywhere in the UK:
- to take a disabled bat, for the sole purpose of tending it and releasing it when no longer disabled, as long as that person can show that it was not disabled unlawfully by them;
- to kill a bat, as long as that person can show that the bat was so seriously disabled, other than by their own unlawful act, that there was no reasonable chance of it recovering.

These defences, however, only apply in circumstances where there is no reasonable alternative, and when the act will not be detrimental to the maintenance of the species at a Favourable Conservation Status (FCS) in its natural range.

Protected areas
Some species of bat found in the UK (greater and lesser horseshoe bats, barbastelle and Bechstein's bat) are listed in Annex II of the Habitats Directive. This means that they can be listed as an interest feature of a SAC and therefore the reason why the SAC is designated. This means they are also a relevant consideration in a Habitats Regulations Assessment (HRA), which provides these species with additional legislative protection. The requirement for this is under Article 6 of the Habitats Directive.[9]

Across the UK Sites of Special Scientific Interest (SSSIs) and Areas of Special Scientific Interest in Northern Ireland (ASSIs) have been identified by the SNCOs. Some such sites have been notified for their bat interest. Legislation relating to such areas

[7] The Wildlife and Countryside Act 1981 has been amended on numerous occasions, in particular by the Countryside and Rights of Way Act 2004 (CROW) and the Natural Environment and Rural Communities Act 2006 (NERC).
[8] Natural England, Natural Resources Wales, Scottish Natural Heritage or Department of the Environment (in Northern Ireland).
[9] A HRA Handbook can be found at http://www.dtapublications.co.uk.

identifies criminal offences if bats are disturbed, if roosts are damaged or if certain operations are undertaken without consent in places notified for their bat interest. In England and Wales the appropriate legislation is the Wildlife and Countryside Act 1981 and in Northern Ireland the Environment (Northern Ireland) Order 2002. In Scotland the Nature Conservation (Scotland) Act 2004 creates and protects SSSIs although no sites have been designated for bats.

Police and court powers

A police constable in any part of the UK has the power, where he has reasonable cause to suspect that a person is committing or has committed an offence, to stop and search them, search or examine any relevant thing in their possession, and seize it. They can also enter land other than a dwelling house without a warrant, or enter and search a dwelling house with a warrant. Constables are empowered to take with them any person or any equipment needed to exercise their powers. Legislation in England and Wales provides a defence for police officers who commit certain offences during the course of their enquiries, otherwise their acts are authorised by a licence issued by the SNCOs.

Those found guilty of offences relating to bats can be sentenced to six month's imprisonment and fined. Recent legislation in England and Wales has removed the maximum amount of fine that can be imposed, and courts there now have the power to impose unlimited fines.[10] In Scotland and Northern Ireland maximum fines at present are set at £5000 but a penalty can be imposed for each animal involved. Courts have a wide range of other sanctions available to them, for example they can order forfeiture of anything used to commit offences or proceeds of crime orders can be made that allow for any profit arising from criminal activity to be confiscated.

Interpretation of legislation

Legislation throughout the UK commonly uses the words: intentional, deliberate or reckless. There is substantial legal opinion as to the meaning of each. Beyond this there have never been any stated cases relating to bats and the criminal law. As such there is little guidance as to the intent of the legislation, with few terms being defined. Commonly questions are posed as to how long bat roosts retain their legal protection when they cease to be used. Some guidance can be found in information produced by the European Union but this information has not been tested in criminal proceedings.[11]

1.2.2 Licensing

The two main types of licence relevant to these guidelines are survey licences (also known as science and education or conservation licences) and EPS licences (also referred to as derogation, mitigation or development licences). Both types of licence permit activities that could otherwise be an offence (see Section 1.2.1).

Survey licences

Survey licences are issued by the following licensing authorities:

- England: NE
- Wales: NRW
- Scotland: Scottish Natural Heritage (SNH)
- Northern Ireland: Department of the Environment (DOE)

These licences do not cover the damage or destruction of a roost site for development; see instead European Protected Species (EPS) licences.

Survey licences are issued to ecologists under the Habitats Regulations to permit them to undertake activities that could otherwise be illegal and lead to an offence, such as entry into a bat roost, temporary disturbance of bats during a survey (including endoscoping) and capture and handling of bats. Ecologists go through a period of training and peer review before being signed off for a licence by their trainer and/or referees. The possession of a survey licence is an indication that the surveyor has reached a minimum standard of training and experience (see Section 2.5.1), although this does not relate to impact assessment or the design and implementation of mitigation, enhancement and monitoring schemes.

Ecologists without a survey licence should not enter known roosts or sites where signs of bat presence (or possible bat presence) have been found. Even where no signs have been found, it is good practice for surveys of potential roost sites to be carried out by ecologists with a survey licence. If it is necessary for an ecologist without a survey licence to survey a building with bat roost potential he/she should immediately withdraw if evidence of bats is found in order for a licensed ecologist to complete the survey. Some surveys, such as emergence or activity surveys, do not require a licence because they do not cause disturbance to bats when undertaken correctly. Some Local Planning Authorities, however, have specific requirements regarding surveyors being licensed if carrying out bat surveys for planning purposes, so local requirements should always be checked.

Although a limited amount of trapping (using mist nets, harp traps and lures) is permitted under some survey licences, a relevant project licence is generally required for such activities and for attaching radio transmitters. Other marking methods, not covered by these guidelines, also require a licence, such as the fitting of tags or rings. A project licence is granted for specific species and numbers of bats, for specific dates and at a particular location. When applying for a project licence, the applicant needs to demonstrate that the level of disturbance is justified and that he or she has the necessary experience to undertake the work.

Conservation licences

Conservation licences may be issued to allow improvements to a bat roost site where the main purpose of the work is for conservation of the species at a specific site. These licences would normally only be issued for a specific proposal at a specific site and only for the duration of the work.

Photography/filming

A licence to photograph (including filming) bats is not required if the photography is an **incidental part of other licensed bat work** and it causes no extra disturbance above that caused by the licensed activities. Such photography includes:

[10] http://www.legislation.gov.uk/uksi/2015/664/contents/made
[11] http://ec.europa.eu/environment/nature/conservation/species/guidance/pdf/guidance_en.pdf

- non-flash photography (i.e. using only natural light or low-level artificial light such as a domestic torch or low-output LED) of roosting bats and of people carrying out licensed work in and around roosts;
- flash photography in roosts and hibernacula only when no bats are present;
- photography of bats caught at traps during survey work;
- flash photography of individual bats for identification purposes or of groups of bats for survey purposes; and
- the use of night vision/infrared/thermal imaging cameras to record roosting (as part of other licensed work) or emerging bats either without the use of further illumination or using infrared illumination (not a red filter).

These only apply where the licence holder considers that this would cause less disturbance than handling or prolonged illumination of bats. It is recommended that there is only one designated photographer at any one time to reduce disturbance.

Flash photography in occupied bat roosts or hibernacula, or entering bat roosts or hibernacula specifically for the purpose of photography (including filming), must be specifically licensed. As disturbing bats specifically for the purpose of photography is potentially very disturbing to bats, licences are only likely to be given where the licensing authority agrees there is a clear need for the photographs and only to experienced photographers who can demonstrate their ability to work efficiently with minimal disturbance to the bats.

Class licences for surveying bats in England

In England a class licensing system has been introduced for survey licences (issued for the purposes of science and education including research). These licences are for all bat-related activities (both voluntary and professional) outside of the NE volunteer bat roost visitor advice service. This includes:

- bat box checks;
- hibernation surveys;
- general survey work;
- professional survey work;
- use of harp traps, mist nets and acoustic lures for development survey purposes.

At present there are four levels of class licence. These are summarised below. The GOV.UK website should be consulted for further details.

- *Level one – to survey bats by observation only (licence WML-CL17) – Disturbance only.*

Surveying of bats by observation only (including the use of artificial light, in the form of torches but not endoscopes) for scientific, research or educational purposes, including informing development projects. This does not include surveys of hibernating bats.

- *Level two – to survey bats using artificial light, endoscopes, hand and hand-held static nets (licence WML-CL18) – Disturbance with handling.*

Surveying of bats using artificial light (e.g. torches), endoscopes, hand and static hand-held nets for scientific, research or educational purposes, including informing development projects. This includes surveys of hibernating bats.

- *Level three – to survey bats using artificial light, endoscopes, hand and hand-held static nets, mist nets and acoustic lures (licence WML-CL19) – Disturbance with handling and mist netting.*

Surveying of bats using artificial light (e.g. torches), endoscopes, hand, static hand-held nets, mist nets and acoustic lures for scientific, research, or educational purposes, including informing development projects.

- *Level four – to survey bats using artificial light, endoscopes, hand and hand-held static nets, harp traps and acoustic lures (licence WML-CL20) – Disturbance with handling and harp trapping.*

Surveying of bats using artificial light (e.g. torches), endoscopes, hand, static hand-held nets, harp traps and acoustic lures for scientific, research, or educational purposes, including informing development projects.

European Protected Species (EPS) licences

EPS licences are issued by the same licensing authorities as survey licences (see previous section). EPS licences are issued under the Habitats Regulations only after three tests have been satisfied in relation to the proposed action, as follows:

- the proposed action must be for the purpose of preserving public health or public safety or other imperative reasons of overriding public interest including those of a social or economic nature and beneficial consequences of primary importance for the environment; and for preventing serious damage to property;
- there is no satisfactory alternative to the proposed action; and
- the action authorised will not be detrimental to the maintenance of the species concerned at a FCS in their natural range.

A FCS is defined in the Habitats Directive as follows (from JNCC, 2007):

> 'i. Population dynamics data on the species concerned indicate that it is maintaining itself on a long-term basis as a viable component of its natural habitats, and
> ii. The natural range of the species is neither being reduced nor is likely to be reduced for the foreseeable future, and
> iii. There is, and will probably continue to be, a sufficiently large habitat to maintain its population on a long-term basis.'

In order for these tests to be correctly applied, it is essential that baseline survey information of a sufficient quantity, quality and standard is supplied. Without this survey information, a licence may not be granted.

Information on when a licence is required, how to apply for a licence, and maintaining the FCS of a species, can be found on the relevant licensing authority websites.

In 2014, Natural England announced the introduction of a Low Impact Bat Class Licence scheme. Ecologists can apply to become a Registered Consultant to use this type of licence, which is for low impact type cases.

1.2.3 Planning policy context

Government policy guidance for biodiversity and nature conservation throughout the UK is provided in the following planning guidance and statements, which are current at the time of writing:

- England:
  - National Planning Policy Framework 2012 (DCLG, 2012)
  - Government Circular 06/2005: Biodiversity and geological conservation – Statutory obligations and their impact within the planning system (DCLG, 2005)
  - Circular 02/99: Environmental impact assessment 1999 (DCLG, 1999)
- Northern Ireland:
  - Planning Policy Statement 2: Natural Heritage (DOENI, Planning Policy Group 2013)
  - Planning Policy Statement 18: Renewable Energy (DOENI, Planning and Environmental Policy Group 2009)
- Scotland:
  - Scottish Planning Policy (Scottish Government, 2014)
- Wales:
  - Planning Policy Wales 2014 (Welsh Government, 2014)
  - Technical Advice Note 5 Nature Conservation and Planning (Welsh Government, 2009)

In addition to the national policy guidance outlined above, regional and local planning policies should be consulted and other country-specific guidance, such as NE's standing advice to Local Planning Authorities (LPAs)[12] may also be relevant.

Government planning policy guidance throughout the UK requires LPAs to take account of the conservation of protected species when considering and determining planning applications. This biodiversity duty is imposed in England and Wales through the Natural Environment and Rural Communities (NERC) Act 2006, which states that 'every public authority must, in exercising its functions, have regard, so far as is consistent with the proper exercise of those functions, to the purpose of conserving biodiversity'. The Nature Conservation (Scotland) Act 2004 states that 'it is the duty of every public body and office-holder, in exercising any functions, to further the conservation of biodiversity so far as is consistent with the proper exercise of those functions'.

Planners are required to consider protected species as a material consideration when assessing a development proposal that, if carried out, would be likely to result in harm to the species or its habitat. This requirement has important implications for bat surveys as it means that, where there is a reasonable likelihood of bats being present and being affected by the development, surveys must be carried out before planning permission is considered.

Adequate surveys are therefore required to establish the presence or absence of bats, to enable a prediction of the likely impact of the proposed development on them and their breeding sites or resting places and, if necessary, to design mitigation, enhancement and monitoring measures.

The term 'development' used in these guidelines includes activities and proposals that could impact bats. In planning terms this includes activities requiring outline and full planning permission but also those that meet the criteria for permitted development, require listed building consent and require prior approval to demolish.

Further details on the standard of information required to assess a planning application is detailed in Clauses 6 and 8 of BS42020 (BSI, 2013). In particular, 'The final report submitted with the application should provide as much certainty as possible and be prepared specifically with the aim of enabling the decision-maker to reach a sound and lawful determination of the application' (Clause 6.3.1)

In addition:

- Clause 7.3 of BS42020 (BSI, 2013) states that 'where an applicant has been advised during pre-application discussions, or have themselves identified that they need to provide information on biodiversity with their planning application, they should ensure that what is submitted is sufficient to enable the decision-maker to validate and register the application'. Preliminary ecological appraisal reports (see Chapter 4) are inadequate to inform the planning process unless no further surveys or mitigation are required.
- The 'Note' with Clause 7.3 of BS42020 (BSI, 2013) states that 'failure to provide all the information required might mean an application is not 'valid' and is not considered or determined'. Therefore, good practice would be for an LPA to include biodiversity in its list of local validation requirements and not to validate an application if bat surveys are required (i.e. if there is a reasonable likelihood that bats could be impacted) but none have been carried out.

Information is also available using the online Bat Planning Protocol.[13]

The planning system should also deliver overall net gains for biodiversity (enhancements), as laid out in the National Planning Policy Framework and other planning policy documents.

1.2.3.1 British Standard for Biodiversity – BS42020:2013

The Code of practice for planning and development set out within BS42020 (BSI, 2013) provides recommendations and guidance for those in the planning, development and land use sectors whose work might affect or have implications for the conservation or enhancement of biodiversity. It aims to:

- promote transparency and consistency in the quality and appropriateness of ecological information submitted with planning applications and applications for other regulatory approvals;
- give planning authorities and other regulatory bodies greater confidence in the information when they consider proposals for development or land management that potentially affects biodiversity;
- encourage proportionality and a good environmental legacy following development.

Further detail can be found on the British Standards Institution website.[14]

[12] https://www.gov.uk/protected-species-and-sites-how-to-review-planning-proposals
[13] http://www.biodiversityplanningtoolkit.com/bats/bio_bats.html
[14] http://shop.bsigroup.com/en/ProductDetail/?pid=000000000030258704

1.2.3.2 Planning trigger list

The planning trigger list in Box 1 presents common development situations where bats are likely to be encountered and therefore where it is most likely that a bat survey will need to be undertaken. **The trigger list is a guide, but it is by no means exhaustive, and professional judgement along with local knowledge should be used to assess where bat surveys are, or are not, appropriate.** Other sites, not listed here, may require a bat survey due to their context, proximity to existing records of bats, the nature of the structure or the proposed activities. Alternative habitats that may initially appear poor for roosting, commuting or foraging bats may be important at particular times of year or in particular situations, for example where other options for bats are limited.

Box 1 Development and planning trigger list for bat surveys, which can be adapted to local circumstances (taken from the Association for Local Government Ecologists (ALGE) template for biodiversity and geological conservation validation checklists 2007, available from http://alge.org.uk/publications/index.php).

(1) Conversion, modification, demolition or removal of buildings (including hotels, schools, hospitals, churches, commercial premises and derelict buildings) which are:

- agricultural buildings (e.g. farmhouses, barns and outbuildings) of traditional brick or stone construction and/or with exposed wooden beams;
- buildings with weather boarding and/or hanging tiles that are within 200m of woodland and/or water;
- pre-1960 detached buildings and structures within 200m of woodland and/or water;
- pre-1914 buildings within 400m of woodland and/or water;
- pre-1914 buildings with gable ends or slate roofs, regardless of location;
- located within, or immediately adjacent to woodland and/or immediately adjacent to water;
- Dutch barns or livestock buildings with a single skin roof and board-and-gap or Yorkshire boarding if, following a preliminary roost assessment, the site appears to be particularly suited to bats.

(2) Development affecting built structures:

- tunnels, mines, kilns, ice-houses, adits, military fortifications, air-raid shelters, cellars and similar underground ducts and structures; unused industrial chimneys that are unlined and brick/stone construction;
- bridge structures, aqueducts and viaducts (especially over water and wet ground).

(3) Floodlighting of:

- churches and listed buildings, green space (e.g. sports pitches) within 50m of woodland, water, field hedgerows or lines of trees with connectivity to woodland or water;
- any building meeting the criteria listed in (1) above.

(4) Felling, removal or lopping of:

- woodland;
- field hedgerows and/or lines of trees with connectivity to woodland or water bodies;
- old and veteran trees that are more than 100 years old;
- mature trees with obvious holes, cracks or cavities, or that are covered with mature ivy (including large dead trees).

(5) Proposals affecting water bodies:

- in or within 200m of rivers, streams, canals, lakes, reed beds or other aquatic habitats.

(6) Proposals located in or immediately adjacent to:

- quarries or gravel pits;
- natural cliff faces and rock outcrops with crevices or caves and swallets.

(7) Proposals for wind farm developments of multiple wind turbines and single wind turbines (depending on the size and location) (NE TIN 051 – undergoing updates at the time of writing).

(8) All proposals in sites where bats are known to be present[1]

This may include proposed development affecting any type of buildings, structures, feature or location.

Notes:

1. Where sites are of international importance to bats, they may be designated as SACs. Developers of large sites 5–10km away from such SACs may be required to undertake a HRA.

Chapter 2

Considerations for bat surveys

2.1 Assessing the need for a bat survey

It is reasonable to request surveys where proposed activities are likely to negatively impact bats and their habitats. However, surveys should always be tailored to the predicted, specific impacts of the proposed activities (see Section 2.2.2). Excessive, speculative surveys are expensive and cause reputational damage to the ecological profession.

Bat surveys may be triggered by a client who wants to purchase land, is in the early stages of designing a project or wants to put in a planning application. Alternatively, a bat survey may be triggered by a LPA that has been advised by an ecologist or used a trigger list or biodiversity checklist (see Section 1.2.3.2) to identify the need for one. Bat surveys may be needed to inform an EPS licence application or a Method Statement to facilitate work being carried out without the need for such a licence. Finally, a bat survey may be triggered after a screening exercise has identified the need for an Environmental Impact Assessment (EIA) or an EIA scoping exercise has identified the need for bat surveys.

2.2 Elements that influence survey design

2.2.1 Stage of proposals

It is good practice for clients to engage with an ecologist as early as possible when planning a project so that ecology can be factored into the design, timetable and budget at an early stage. Later engagement can result in late design changes and extra delays and costs.

In addition to the client engaging with an ecologist, early engagement with the LPA and the relevant licensing authority is also beneficial. These two bodies have different functions and may take different decisions on the same proposal. In addition, the granting of, or lack of need for, planning permission does not negate the need to consider protected species legislation.

It is necessary to know what stage the project is at in order to design surveys according to the amount of detail that is required, for example a client considering the purchase of land is likely to require less detail than is required for an EPS licence application and surveys will need to be tailored accordingly.

Large projects such as road schemes or power stations often commence years before any work is carried out on the ground and so surveys in the early years of the project may be at a broad level to identify features of high conservation value to inform project design, with more detail gained later on. It may also be necessary to repeat surveys on projects with long lifespans so that survey data remains current (see Section 2.6.3 for considerations with respect to age of survey data).

2.2.2 Potential impacts

The purpose of professional surveys is generally to carry out an assessment of the impacts likely to arise from proposed activities. An ecologist should be provided with (or request) enough information about a project from the start to identify the likely ecological impacts (or lack of impacts) from an early stage. These should be reviewed throughout the project, particularly on larger projects where the proposals may be subject to change over time.

Some impacts on bats and their habitats that can arise from proposed activities are given in Table 2.1.

Table 2.1 Impacts on bats that can arise from proposed activities

Impacts on... ...bats	...roosting habitats	...commuting and foraging habitats
Physical disturbance Noise disturbance through, for example, increased human presence or use of noise-generating equipment Lighting disturbance Injury/mortality (e.g. in roost during destruction or through collision with road/rail traffic)	Modification of access point to roost either physically or through, for example, lighting or removal of vegetation Modification of roost either physically, for example by roof removal, or through, for example, changed temperature, humidity, ventilation or lighting regime Loss of roost	Modification of commuting or foraging habitats either physically or through disturbance, e.g. light spill/noise Severance of commuting routes (fragmentation) Loss of foraging habitats

Different parameters to consider when assessing the different impacts of a project are:

- Is it a positive or a negative impact?
- What is the extent of the impact? What area does it cover?
- What is the magnitude or size of the impact?
- What is the duration of the impact? How long will it last?
- What is the timing and frequency of the impact?
- Is the impact reversible? Will it be temporary or permanent?
- How do the impacts differ throughout the process from pre-construction through construction to operation (and dismantling and restoration for some projects).

More information can be found in *Guidelines for Ecological Impact Assessment in the UK and Ireland: Terrestrial, Freshwater and Coastal* (CIEEM, 2016).

The unique combination of project and site will influence the type and nature of potential impacts that are relevant to different projects. Understanding how these elements work together is the key to good survey design.

2.2.3 Zone of influence and defining the survey area

A client should provide a plan showing the site boundary (or red-line boundary for planning purposes), which indicates the area within which proposed activities will take place. Predicted impacts within this boundary will influence the spatial design of surveys. Other considerations when defining survey area are given below.

- The 'zone of influence' of the proposed activities may be different from the site boundary. The term zone of influence (ZoI) is used in formal EIA projects (although the principle can be applied to any project) and is defined by CIEEM (2016) as 'the areas/resources that may be affected by the biophysical changes caused by activities associated with a project'.
- The client's land ownership (the blue line boundary for planning purposes) will determine where access for surveys may be easily obtained.

All ecologists working on the project should understand how the survey area has been defined and the definition should be revisited as the project evolves. It is essential for an ecologist to be familiar with up-to-date plans and review the surveys that have been, and will be, carried out accordingly.

2.2.4 Defining aims and objectives

It is important at the start of any survey that the **aims and objectives are clearly defined and that the survey report subsequently demonstrates how these have been met.**

The aims of surveying at a proposed development site are generally to:

- collect robust data following good practice guidelines to allow an assessment of the potential impacts of the proposed development on bat populations both on and off site;
- facilitate the design of mitigation, enhancement and monitoring strategies for bats;
- provide baseline information with which the results of post-construction monitoring can be compared, where appropriate;
- provide clear information to enable the LPA and licensing authority to reach a robust decision with definitive required outcomes;
- assist clients in meeting their statutory obligations; and
- facilitate the conservation of bat populations.

Early objectives in a project may be to:

- establish what stage the project is at and therefore what action is needed;
- define the survey area; and
- carry out a preliminary ecological appraisal for bats (Chapter 4) or preliminary roost assessment (Chapters 5 and 6) to inform the design of subsequent, more detailed surveys.

Later objectives may be to:

- obtain roost count data during one active period; and
- trap bats to identify to species level and gain information on gender and breeding status.

Aims and objectives should be revisited throughout a project because each stage of surveying informs the next; bat surveys are an iterative process, which should not usually be fixed from the outset.

2.2.5 Proportionality

When planning surveys **it is important to take a proportionate approach**. The type of survey (or suite of surveys) undertaken and the amount of effort expended should be proportionate to the predicted impacts of the proposed activities on bats. Clause 4.1.2 of BS42020 (BSI, 2013) states that 'professionals should take a proportionate approach to ensure that the provision of information with the (planning) application is appropriate to the environmental risk associated with the development and its location'.

Below are other elements that influence the type of survey and effort expended, the examples given being descriptive rather than prescriptive:

- Likelihood of bats being present (e.g. it is often harder and thus may require more survey effort to show that bats are, on the balance of probability, absent from structures rather than present. However, once presence has been established, further surveys may be required to characterise the roost).
- Type of proposed activities (e.g. targeted survey effort may be required for project types known to have specific impacts such as a road scheme or wind farm).
- Scale of proposed activities.
- Size, nature and complexity of the site.
- Species concerned (e.g. some species are harder to detect using standard techniques (such as Bechstein's bat) or are of particular conservation importance (e.g. Annex II species). Different survey types and more survey effort may be necessary if the site is within the range of such species and habitats on site are suitable for the species).
- Numbers of individuals (e.g. sites with larger numbers of individuals (maternity or hibernation roosts or key commuting routes and foraging areas) may require more survey effort to establish numbers or species assemblages).

2.2.6 Considering data analysis

Where large amounts of bat activity data are collected using static/automated bat detectors (see Section 8.2) or radiotelemetry is used (see Section 9.3), statistical analysis is important because the meaning is not readily understood just by looking at the data. In particular, trapping and radiotelemetry surveys are highly intrusive and can have implications for bat welfare so a clear plan of why the data is needed, what data is to be collected and how the data will be analysed is essential. If the methods of analysis (see Sections 10.3 and 10.4) are chosen at the survey design stage, this ensures that such testing is possible and makes testing much easier. Data analysis should be an integral part of such surveys and if data collection and analysis are not standard then consideration should be given to conducting a pilot survey.

It is essential that data collected for direct comparison has been collected in the same way, and ideally by the same equipment (e.g. bat detectors, which should be subject to regular testing and calibration); and in suitable conditions, otherwise these factors can introduce bias – differences detected may relate to these factors rather than to real differences on the ground.

In addition, the term 'bat pass' could have a different definition according to equipment and operator, therefore it is important to be clear on how 'bat pass' will be defined when setting out. Some ecologists use 'bat pulse' as the unit of bat activity (rather than bat pass) when analysing their data (see Sowler and Middleton, 2013). The important point is to be consistent.

The main message is that there are various elements that can add bias to survey results and this bias should be minimised as far as is practical.

2.2.7 Mitigation hierarchy

The mitigation hierarchy dictates that **impacts should be avoided in the first instance but, where impacts cannot be avoided, then they should be adequately mitigated or, as a last resort, compensated for** (refer to National Planning Policy Guidance para 118, DCLG, 2012). Where mitigation is referred to in these guidelines it should be taken to mean all the elements of the mitigation hierarchy. Definitions of these terms are provided in the *Bat Mitigation Guidelines* (Mitchell-Jones, 2004) and reproduced below.

Mitigation: in this strict sense, mitigation refers to practices which reduce or remove damage (e.g. by changing the layout of a scheme, or altering the timing of the work).

Compensation: this refers to works which offset the damage caused by activities (e.g. by the creation of new roosts).

Following the preliminary ecological appraisal (see Chapter 4) or preliminary roost assessment (see Chapters 5 and 6) it may be possible to identify potential impacts and adjust the design or timing of the project to avoid them. The extent to which impacts can be avoided will influence the design of further surveys. In some circumstances, further surveys may not be needed, in others it may be necessary to collect baseline information against which to compare monitoring data to assess whether impact avoidance has been successfully implemented.

Where negative impacts cannot be avoided through design ('embedded mitigation'), it is reasonable to recommend further bat surveys to facilitate an impact assessment and design a mitigation and monitoring strategy.

2.2.8 Using good practice guidance

BS42020 (BSI, 2013) states, in relation to reports submitted with planning applications (although the same principles apply to reports produced as part of an EPS licence application or for other purposes):

> Methods used to undertake surveys and to prepare information presented in ecological reports should (except in the circumstances described below) follow published good practice guidelines where they exist. Claims of compliance with good practice should be substantiated (Clause 6.3.6).
>
> A competent ecologist should, as appropriate, modify their approach from that of published good practice or standing advice issued by a statutory body where, for example:
> (a) it is necessary to adapt to the specific requirements of a case or site;
> (b) an innovative approach might improve upon published good practice and/or provide a more valuable outcome;
> (c) it might only be appropriate to follow good practice guidance in part as the guidance offers a range of optional methods (e.g. for surveys), of which only one is appropriate to the study in question; or
> (d) published good practice is out of date and/or where better techniques have been developed and recognised throughout the profession (Clause 6.3.7).

2.3 Bat surveys for development

Figure 2.1 illustrates the process that ecologists should go through when carrying out professional bat survey work where activities are proposed that could impact bats.

Figure 2.1 The process of carrying out professional bat surveys for proposed activities that could impact bats.

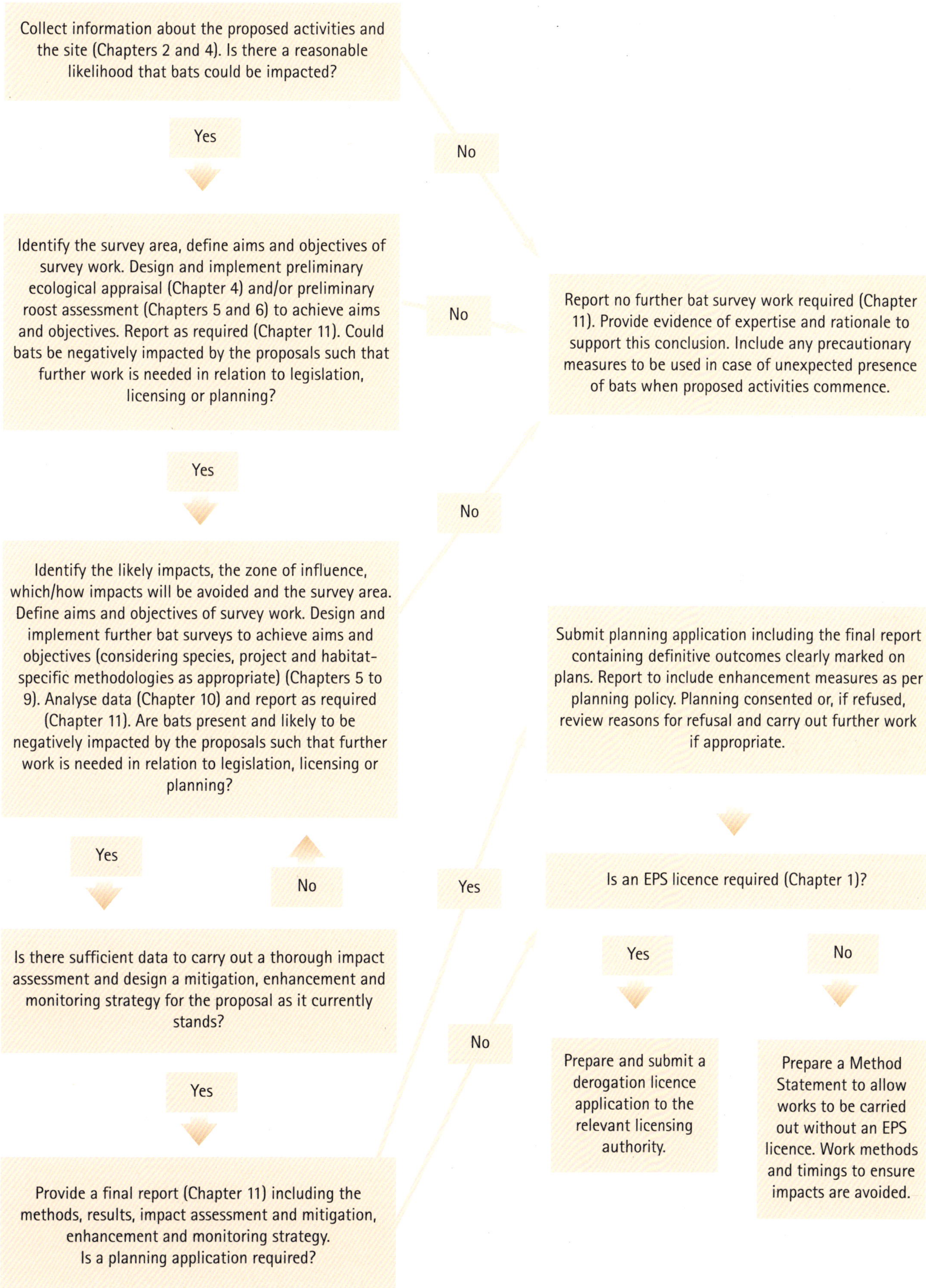

2.4 Survey timing

Bats use different roosts, commuting routes and foraging areas throughout the year according to their life cycle (see Section 3.2) and the availability of their insect prey, which are both influenced by the ambient conditions (temperature, humidity, rainfall, wind) at the location in question. Multiple surveys are usually needed to investigate temporal or seasonal changes in activity; readers should refer to the individual survey chapters (Chapters 4 to 9) for more information. For landscape-scale or higher-impact projects, it is often appropriate to collect data for at least a year.

Table 2.2 provides optimal timings for all types of survey described in these guidelines, although individual survey chapters (Chapters 4 to 9) provide further clarification/caveats with respect to timings. **An experienced surveyor should carry out surveys at a time that gives them the highest chance of establishing whether or not bats are present and how they are using the habitat (including roosts). Actual timings will depend on a number of factors including the surveyor's knowledge and experience of the site and surrounding habitats, existing data records, possible bat species present, geographical location, weather conditions in that particular year and, of course, the aims and objectives of the survey.**

Table 2.2 Recommended UK survey times for survey types described in these guidelines.

Survey type	Month											
	J	F	M	A	M	J	J	A	S	O	N	D
Preliminary ecological appraisal - fieldwork	optimal	optimal	optimal	optimal	optimal	optimal	optimal	optimal	optimal	optimal	optimal	optimal
Preliminary roost assessment – structures[a]	optimal	optimal	optimal	optimal	optimal	optimal	optimal	optimal	optimal	optimal	optimal	optimal
Emergence/re-entry survey for maternity or summer roosts[b]					optimal	optimal	optimal	optimal				
Emergence/re-entry[c] survey for transitional roosts[b]				weather/location dependent					weather/location dependent	weather/location dependent		
Emergence survey for mating roosts[b]								optimal	weather/location dependent	weather/location dependent		
Hibernation survey – structures[a]	optimal	optimal	weather/location dependent								weather/location dependent	optimal
Preliminary ground level roost assessment - trees[d]	optimal	optimal	optimal	sub-optimal	sub-optimal	sub-optimal	sub-optimal	sub-optimal	sub-optimal	sub-optimal	sub-optimal	optimal
Potential roost feature (PRF) inspection survey - trees	optimal	optimal	optimal	optimal	optimal	optimal	optimal	optimal	optimal	optimal	optimal	optimal
Ground level bat activity survey – transects and automated/static				weather/location dependent	optimal	optimal	optimal	optimal	weather/location dependent	weather/location dependent		
Pre-, during and post-hibernation – automated/static bat activity survey	weather/location dependent	weather/location dependent	weather/location dependent								weather/location dependent	weather/location dependent
Swarming survey								optimal (second half)	optimal	optimal		
Back-tracking survey				weather/location dependent	optimal	optimal	optimal	optimal	weather/location dependent	weather/location dependent		
Trapping survey[e]					optimal	sub-optimal	sub-optimal (first half), optimal (second half)	optimal	weather/location dependent	weather/location dependent		
Radio tagging and tracking survey[e]					optimal	sub-optimal	sub-optimal (first half), optimal (second half)	optimal	weather/location dependent	weather/location dependent		

[optimal shading] = optimal period

[sub-optimal shading] = sub-optimal period

[weather or location dependent shading] = weather or location dependent (i.e. may not be suitable due to spring and autumn conditions in any one year or in more northerly latitudes). Note that October surveys are not acceptable in Scotland.

[a] Not including trees

[b] Please see: Table 7.1 (page 51) for recommended timings for surveys to give confidence in a negative result. For sites assessed as having low suitability a survey should be carried out between May and August. For sites with moderate and high suitability a proportion of the surveys should be carried out between May and August (to detect maternity roosts if present) but some of the surveys may be carried out later in the year in order to detect transitional and mating roosts. The survey season for presence/absence surveys is defined as May to September. Roost characterisation surveys may be appropriate in April and/or October depending on the need to characterise transitional roosts at these times.

[c] The use of dawn surveys in autumn should be clearly justified because longer nights and poorer weather conditions may result in bats returning to roosts early and not re-emerging for pre-dawn foraging, producing a false negative survey result.

[d] Tree surveys can be sub-optimal in the spring, summer and autumn due to foliage obscuring parts of the tree. If all parts of the tree are visible then the survey can be carried out at any time. If parts of the tree are obscured by foliage then it is not possible to carry out a thorough survey and this limitation should be recognised and the impact on the results acknowledged. Please refer to Chapter 6 for more information.

[e] Trapping and tagging surveys should avoid the time when bats are heavily pregnant or lactating unless a project specific licence allows such activities, based on the information needs of the project. Please refer to Chapter 9 for more information.

2.5 Resources for surveys

2.5.1 Human resources

It is important for those commissioning, scheduling, undertaking and assessing bat survey work to ensure that the ecologists carrying out the work have sufficient training, skills, experience and licences. There is a multitude of bat survey types and the equipment required to carry them out is technical and varied. **None of these surveys can be carried out effectively without specific training and some work also requires ecologists to hold licences to carry out the work legally** (see Section 1.2.2). Alongside survey skills, ecologists planning surveys, leading survey teams, carrying out impact assessments and designing mitigation, enhancement and monitoring schemes require a whole suite of other knowledge and expertise. It is the responsibility of the ecologist and their employer to ensure that appropriate training, skills, experience and licences are in place before carrying out a survey.

Clauses 4.4.1 and 4.3.2 of BS42020 (BSI, 2013) state that 'development proposals that are likely to affect biodiversity should be informed by expert advice' and that ecologists 'should only attempt to offer a bone fide ecological opinion if they have the necessary knowledge, skills and experience to do so, or have secured appropriate competent assistance' respectively.

Clause 4.3.4 of BS42020 (BSI, 2013) states that 'evidence of qualifications, additional training and experience should always be available on request as further evidence of an individual's competence in a particular field of knowledge or area of expertise'.

Training and experience can be gained through mentoring by an experienced and licensed ecologist or attending training courses run by organisations such as the BCT, the Chartered Institute of Ecology and Environmental Management (CIEEM) or other private providers. Local bat groups (LBGs) can also provide training, although this is generally aimed at those carrying out voluntary bat work, for which the aims of surveys are likely to be different. Although skills such as handling and bat identification remain the same for both types of surveys, additional knowledge, skills and experience (such as the ability to design surveys, lead survey teams, assess impacts and design mitigation, enhancement and monitoring strategies) are required to carry out bat surveys professionally.

The BCT published *Professional Training Standards for Ecological Consultants* in 2012. This describes the knowledge and skills required to competently undertake professional bat work to three experience levels, which are described below.

- Level One: To independently and competently undertake professional surveys involving bats.
- Level Two: In addition to the above, to independently and competently lead survey teams/design surveys, assess impacts and design mitigation.
- Level Three: In addition to the above, to independently and competently undertake advanced survey techniques (e.g. trapping and attaching radio transmitters).

The professional training standards document (Bat Conservation Trust, 2012) describes the knowledge and understanding/skill and experience requirements for different topic areas (e.g. Unit 1 Foundation knowledge, Unit 2 Legislation, licensing and planning, Unit 3 Preparation and planning of surveys) in relation to the levels described above and provides performance criteria against which these can be assessed.

CIEEM published *Competencies for Species Survey: Bats* in 2013 (CIEEM, 2013a) in association with BCT, which also describes knowledge, skills and experience required to carry out professional bat work.

While membership of a professional body such as CIEEM (or Chartered Ecologist or Environmentalist status) does not provide evidence for a skill level with respect to bats or other species, members are required to conform to a Code of Professional Conduct. CIEEM's Code of Professional Conduct states that members will:

(i) maintain and develop their professional knowledge and skills and work normally within their sphere of competence; and
(ii) seek advice and assistance if they are involved in topics outwith their sphere of competence (CIEEM, 2013b).

The CIEEM website hosts a professional directory of members,[15] which can be searched according to the services provided.

2.5.2 Equipment, documentation and data recording

The documentation/equipment chosen for a survey should make the survey safer, easier and more efficient, thorough and accurate. Requirements for documentation/equipment depend on the nature of the survey, the nature of the site and factors such as the client/owner's health and safety requirements. Lists of equipment relevant to different survey types are provided in Appendix 1. A generic list of both documentation and equipment appropriate to all field surveys for bats is provided below:

- any documents that are necessary to allow approved access to the site;
- risk assessment (and biosecurity risk assessment as appropriate);
- any other health and safety documentation;
- copies of relevant licences for the survey activities;
- maps/aerial photographs of the site and surrounding area;
- maps/plans/drawings of site features, clearly illustrating the site boundary;
- any previous survey or background information;
- survey form or digital equipment suitable for recording such as a smartphone, tablet or GPS recorder;
- digital camera;
- spare batteries, bulbs and memory cards for all equipment;
- personal protective equipment (PPE, for example steel toe-capped boots, hard hat, overalls, high visibility jacket, gloves, dust mask);
- first aid kit; and
- charged mobile phone (ensure there is network availability at the site in question and ensure back-up such as hand-held radios or buddy system if no signal).

[15] http://www.cieem.net/members-directory

Where it is necessary to use technical measuring devices (e.g. a thermometer) or recording equipment (e.g. a bat detector), it is essential that the equipment is both calibrated and tested on a regular basis to ensure that when the results are compared this is a like-for-like comparison. Similarly, it is essential to have a good understanding of the settings and the sampling rate of detectors. Different bat detector microphones vary in their sensitivity (Adams *et al.*, 2012) and this should be considered. The benefit of recording bat activity is that there is an auditable record of work carried out; data should be retained for this purpose.

Data recorded during a survey should be accurate, thorough and consistent across surveys of the same type. Standard survey forms should be used for each survey type to prompt the ecologist to record all the information necessary (and no more) and allow the raw data to be passed on if the need arises, such as in a public inquiry.

When recording survey results, it is obviously important to record positive sightings but it is also important to make a record where a site or feature has been surveyed but returned a negative survey result (i.e. not suitable for bats or no evidence of bats found). This information can be just as important when justifying subsequent actions undertaken.

2.6 Dealing with survey limitations

Clause 6.7.1 of BS42020 (BSI, 2013) states that 'To reduce uncertainty, and to achieve full scientific disclosure, those undertaking surveys and preparing ecological advice and reports should identify all relevant limitations' with respect to methods and site conditions. Clause 6.7.2 of BS42020 (BSI, 2013) states that 'any limitations associated with work should be stated, with an explanation of their significance and any attempt made to overcome them. The consequences of any such limitations on the soundness of the main findings and recommendations in the report should be made clear.'

2.6.1 Weather conditions

The weather affects bat activity and therefore surveyors should check weather forecasts prior to surveys for active bats and record weather conditions, **including temperature, wind speed and precipitation**. These variables should be recorded at the start and end of each survey and if conditions change during the survey. When ecologists are not present (for example, during automated/static monitoring surveys) options for recording weather conditions include a temperature logger, a weather recording station and obtaining meteorology data online. This data provides context to the survey results and therefore a plan should be in place to ensure it is recorded/obtained.

Additionally, the weather conditions prior to the survey may influence bat activity (e.g. a dry spell after a long period of rain may result in bats foraging for longer because they are hungry) and could be recorded and reported if this is the case.

The effect of weather conditions on active bats is likely to be different for different species (with different flight capabilities) in different situations (for example, open versus sheltered habitats).[16]

The aim should be to carry out surveys in conditions that are close to optimal (sunset temperature 10°C or above, no rain or strong wind), particularly where only one survey is planned. Where multiple surveys are planned, carrying them all out in optimal conditions enables a like-for-like comparison of results, although it is recognised that in spring and autumn, and particularly in more northerly latitudes, these conditions may be rarer and some of the surveys may need to be carried out at lower temperatures or in more windy conditions. This situation does provide some insight into how the bats respond to poorer conditions. Surveys carried out when the temperature at sunset is below 10°C should be justified by the ecologist and the effect on bat behaviour considered. In cooler, wetter and windier conditions bats may not emerge, emerge later, forage for shorter time periods, carry out fewer foraging bouts or use alternative, more sheltered habitats.

2.6.2 Restricted access

Clients may have specific requirements for access to sites such as items of PPE, documentation or that surveyors are escorted by site personnel. Some sites may require specialist equipment; for example, gas monitors in a confined space. Site-specific requirements should be established before the site visit and should not be cited as limitations to a survey if they can be met through advance planning.

Sometimes it is not possible to gain permission to access land. In this situation, it is recommended that a record of access requests and any responses received are retained as evidence that access permission was sought but was not granted.

Access to survey may also be restricted for health and safety reasons; for example, a building may be structurally unsound or a tree may not be safe to climb. Documentation may be available from a structural engineer or arboriculturist as evidence but, if not, justification should be provided in the bat survey report). In situations where a thorough preliminary ground level roost assessment or PRF inspection survey is not possible, the number of presence/absence surveys may need to be increased accordingly.

The impact of any remaining limitations (relating to access) on the resulting data should be acknowledged in the report.

2.6.3 Age of survey data

Ideally, the survey data should be from the last survey season before a planning or licence application is submitted, although often data older than this can have considerable value, particularly where collected over a number of years using different techniques. The value of older data should be considered when updating surveys as it may not be necessary to start from scratch.

[16] Kronwitter (1988) studied the influence of temperature and precipitation on the activity of noctule bats in Germany, observing no emergence, late emergence and fewer foraging bouts in cooler conditions and later emergence in rainy conditions. Slack and Tinsley (2015) looked at bat activity at wind farm sites and found no bat activity at temperatures below 6°C, limited bat activity below 10°C and a reduction in bat activity at wind speeds of 5.4m/s and greater. Radio-tagged barbastelle bats exhibited the same behaviour in wind speeds of 11m/s as on previous calmer nights in a study by Davidson-Watts (2014a).

In some cases, data may be needed from the night before operations are carried out either to confirm that bats have left an identified roost, or as a precautionary measure.

The length of time survey data remains valid should be decided on a case-by-case basis and is dependent on a number of questions, as follows:

- Were the original surveys carried out according to good practice guidelines?
- Were the original surveys constrained in any way (in terms of timings, weather conditions, equipment used, number of surveyors, surveyor expertise, etc.)?
- Do the results of the original surveys support the original conclusions and are these still relevant?
- Has the nature of the site or the surrounding area changed since the original surveys (e.g. has a structure deteriorated and become less suitable for a roost or has human occupation ceased and the structure become more suitable for a roost)?
- Are additional surveys likely to provide information that is material to a decision (such as a planning consent), the design of mitigation measures, or specific advice relating to a proposed activity?

2.6.4 Other potential limitations

The availability or cost of equipment should not be cited as a reason for not using the most appropriate piece of equipment for a bat survey. Professional ecologists should ensure that they consult with the client to establish the nature of the site and scrutinise bat records and previous survey results to ensure they have the right equipment to carry out their work.

Some equipment is inherently constrained but still the most appropriate equipment for the job; for example, bat detectors can only provide an index of activity rather than absolute numbers of bats (see Section 10.2) and some species are difficult to detect due to their quiet echolocation calls (see Section 3.9). These constraints should be acknowledged in the report and methods to overcome them described.

Bat surveys are seasonally constrained and this should be factored into project scheduling to ensure that surveys are carried out at the most appropriate time of year. Ideally, timing should not be cited as a limitation to the survey.

2.7 Health and safety

It is the legal duty of an employer to have a written health and safety policy unless they employ fewer than five employees (although even in this situation it is good practice to have a policy in place). Guidance on safety and risk management can be found on the HSE's website.[17]

A hazard is something that has the potential to cause harm; it is associated with a degree of danger and is quantifiable in terms of its severity. Risk is the actual likelihood of harm from a particular hazard. If a risk is considered too high then the proposed action should not be undertaken or measures should be applied to either remove the hazard or avoid/reduce the risk that the hazard poses. It is generally more appropriate for bat surveys to be undertaken in pairs or within a larger team due to the potential risks involved. However, it may be possible to adequately control the risks to a lone worker in certain circumstances.

Bat surveys have some very specific risks arising from particular hazards such as working at height, confined spaces, asbestos and night-time working resulting in tiredness. It is important that these hazards are adequately considered and risks are adequately controlled before surveys are undertaken. The most effective way to ensure this is by carrying out a risk assessment. A targeted risk assessment should be prepared and completed for every site, to ensure that any site-specific risks are considered alongside generic risks. On arrival on site, for every visit, the risk assessment should be reviewed to establish that all possible risks have been taken into account. There should be a mechanism in place for items to be added to the risk assessment and for this information to be available for subsequent site visits (particularly important if different staff are deployed each time).

Appendix 3 lists hazards and risks associated with bat fieldwork and measures that can be taken or equipment that can be used to manage them. Sample risk assessments and guidance on completing them can be found on the HSE website.[18] Guidance on carrying out risk assessments for lone working is also available from the Member's Area of the CIEEM website.[19] General guidance on health and safety is also provided in CIEEM's *Good Working Practices* (CIEEM, 2013c).

In some situations, particularly for larger developments, the site owner/developer/client will also have their own risk assessment, a health and safety induction and/or other related procedures.

All equipment used should be regularly checked and maintained, in line with appropriate legislation (this may require formal inspections by accredited bodies).

The following types of work require advanced knowledge and the use of specialist equipment; information can be gained on the specialist training courses indicated.

- Work in confined spaces (tunnels, culverts, etc.) – confined spaces training course.
- Working at height – working at height training courses provide training on the safe use of ladders and assessment of which equipment is appropriate to the task.
- Work in trees – arboricultural climbing course provides training in the use of specialist equipment and climbing/aerial rescue techniques.
- Work underground (mines, caves, etc.) – confined spaces training course, mine safety course. Basic caver training and advice on safety issues in specific local caves and mines can also be obtained from the British Caving Association (BCA), Regional Caving Councils or local caving clubs.
- Work on a construction site – to get an Ecologist CSCS (Construction Site Certification Scheme) card, you need to apply for the card through the BALI (British Association of

[17] http://www.hse.gov.uk/pubns/indg449.pdf
[18] http://www.hse.gov.uk
[19] http://www.cieem.net

Landscape Industries) website.[20] Before you can apply you need to attend a 1 day ROLO H&S training day and sit the touch screen test.
- Work in buildings which may contain asbestos – asbestos awareness training course. Asbestos may be present in structures built before 2000; some such buildings may have an asbestos risk register that can be requested and scrutinised prior to entry.

Whether employers provide vehicles or expect employees to drive their own for work purposes, they should have a policy to address working hours, time spent driving and vehicle maintenance. The Health and Safety Executive (HSE) estimates that up to a third of all road traffic accidents involve a driver who is at work at the time. Road accidents are a particular risk for ecologists carrying out nocturnal bat surveys, as the functionality of a driver decreases with increasing sleep deprivation or fatigue. Companies therefore have a duty to develop policies to ensure safe working practices, and it is recommended that driving is included in working hours in these policies.

Where phone reception is poor it may be necessary to use 112 in an emergency.[21]

2.8 Insurance

Before undertaking any work for a client, ecologists should have appropriate insurance, including professional indemnity insurance and public liability insurance. For members of CIEEM, adequate insurance cover is a strict requirement of membership.

Professional indemnity insurance can help protect an ecologist if claims are brought against him or her by a client, due to a perceived problem with the work undertaken. Professional indemnity insurance is needed if an ecologist provides advice to a client, handles data belonging to a client, is responsible for a client's intellectual property, or provides professional services, and if an ecologist's work could be challenged or questioned. Ecologists may be vulnerable to claims of negligence if professional advice or services fail to meet a client's expectations or are perceived to cause financial loss.

Public liability insurance covers the compensation an ecologist may have to pay a client, contractor or member of the public, due to accidental injury or property damage caused by the ecologist either on the ecologist's premises, during field surveys or at a client's premises.

2.9 Summary

Ecologists should be considering the following questions as they carry out their professional survey work:
- Is there a need for survey work to be carried out?
- Is the purpose of this work understood in relation to the current stage of the project?
- Have the aims and objectives of the work been clearly defined and are these fit for the purpose they were intended?
- Will the stated aims and objectives of the survey work be achieved?
- Is the survey work proportionate to the impacts?
- Have the potential impacts, the ZoI and the impacts that could be avoided through design been adequately assessed?
- Is the defined survey area appropriate?
- Are the most appropriate survey types being used?
- Are the surveys being carried out according to good practice? If not, then how will any limitations be accounted for?
- Do the surveys fit in with the planned project schedule? Do the surveys or schedule need to be amended?
- Does the team have the competence to carry out the survey work?
- Does the team have the capacity to carry out the survey work?
- Has the right equipment been chosen for the survey work? Does the team have the right equipment? Does the equipment need calibrating, testing or servicing?
- Is all of the appropriate data being recorded?
- Are there any specific health and safety requirements that need to be fulfilled and will this impact on the survey results/survey schedule?
- Is site access available to allow the surveys to be carried out efficiently and effectively within the defined survey area?
- Has the project been altered recently such that the surveys or schedule need to be reviewed?
- Has all the relevant information been requested from the client and communicated back?
- Have clear and definitively stated outcomes been provided to enable the LPA to include conditions in a planning decision?
- Have the client's expectations been realistically managed in terms of meeting good practice and being clear on planning and licensing requirements?

[20] http://www.bali.org.uk/quality_assurance/liss_cscs/ occupations#environmental
[21] See https://www.youtube.com/watch?v=XPZv_8dABfU for information on maximising chances of getting hold of the emergency services.

Chapter 3

Ecological considerations for bat surveys

3.1 Introduction

Knowing about the ecology of the different species is necessary to determine how bats are likely to use the landscape, so that appropriate survey methods can be chosen. Bats are cryptic, use large geographical areas in three-dimensional space, and have the potential to disperse over large areas, so that they are difficult to survey without an understanding of their ecology.

Due to the relatively recent discovery of Alcathoe bat (*Myotis alcathoe*), the ecology of this species in the UK is poorly understood and therefore this species is not included in the species tables in the following sections.

3.2 Bat life cycle

Figure 3.1 provides a visual representation of the life cycle of a bat; further descriptions are provided in the text below.

Figure 3.1 Bat life cycle.

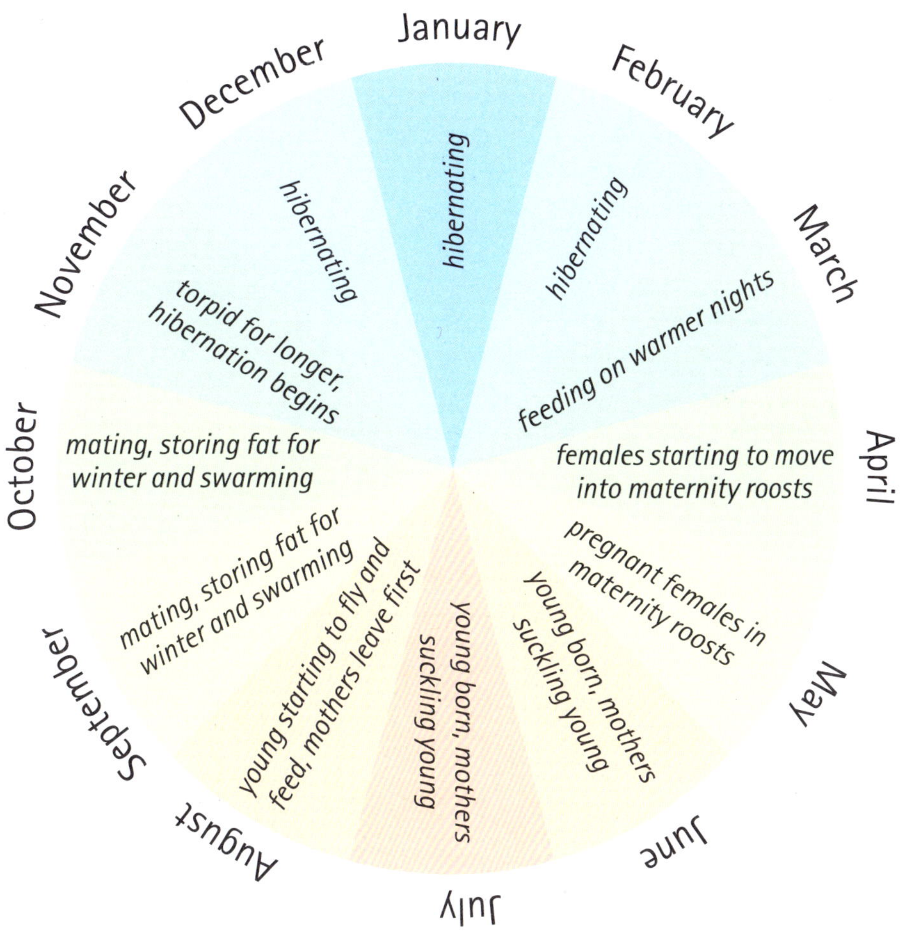

UK bats spend much of the winter (dependent on conditions in any one year at any specific location) in torpor at hibernation sites, although they will rouse on warmer nights to drink, forage and expel waste products. Bats can also change hibernacula depending on weather conditions.

During the spring bats feed more and more during the night and the period from April (likely to be slightly later in northerly latitudes) to early June is a time of intense feeding to recover weight lost during the winter. During this time, females gather together at maternity roosts that provide appropriate conditions to rear young. In some species, males are also present in maternity roosts although for many species the males roost elsewhere either individually or in small groups.

Birthing times can be highly variable between locations, years, species and even between individuals of the same species. However, the main period for births is June, then the young begin to fly in July and August, at first still taking milk from their mothers but gradually becoming more independent (Dietz *et al.*, 2011). As the young become independent, the females disperse to find mates and gain weight before winter.

During autumn, many *Myotis* bats swarm at caves and mines to mate and/or find a hibernation site. Males of some species establish mating territories where they may fly or call specifically to attract females.

As the weather turns colder, bat activity declines and foraging becomes restricted to warmer nights. Bats spend progressively more time in torpor and slowly return to their hibernacula.

3.3 Bat roost types

The definitions of different roost types in Table 3.1 have been taken from the NE EPS licence application form available at the time of writing.

Table 3.1 Bat roost types (from NE EPS licence application form available at the time of writing).[a]

Roost type	NE definition
Day roost	A place where individual bats, or small groups of males, rest or shelter in the day but are rarely found by night in the summer.
Night roost	A place where bats rest or shelter in the night but are rarely found in the day. May be used by a single individual on occasion or it could be used regularly by the whole colony.
Feeding roost	A place where individual bats or a few individuals rest or feed during the night but are rarely present by day.
Transitional/occasional roost	Used by a few individuals or occasionally small groups for generally short periods of time on waking from hibernation or in the period prior to hibernation.
Swarming site[b]	Where large numbers of males and females gather during late summer to autumn. Appear to be important mating sites.
Mating sites[c]	Where mating takes place from late summer and can continue through winter.
Maternity roost[d]	Where female bats give birth and raise their young to independence.
Hibernation roost	Where bats may be found individually or together during winter. They have a constant cool temperature and high humidity.
Satellite roost	An alternative roost found in close proximity to the main nursery colony used by a few individual breeding females to small groups of breeding females throughout the breeding season.

[a] The table defines roost types for the purposes of consistency but it should be noted that not all of these sites are also breeding sites, resting places or places used for shelter or protection as described in the legislation. Judgements as to what is protected under law should be undertaken on a case-by-case basis (the term roost is not used in the legislation). The EU has provided guidance on this point in: Guidance on the strict protection of animal species of community interest (2007). Please also see Sections 1.2.1 and 1.2.2.

[b] Roosting may occur alongside the swarming activity and it is the structures used for rest and shelter within the swarming site that are the roost.

[c] Mating sites can include those where bats call for mates on the wing; however, these are also associated with a place that the mating takes place, which is the mating or harem roost.

[d] In some species, males may also be present in the maternity roost.

3.4 Species roosting preferences

Table 3.2 provides information from studies of the roosting preferences of different bat species. It should be noted that this table is not exhaustive and was not derived from a thorough literature search – species may be found to roost in different locations to those described here.

Table 3.2 Roosting preferences of different species.

Species common name	Species scientific name	Roosting preferences
Greater horseshoe bat	*Rhinolophus ferrumequinum*	During the summer females use large, old, undisturbed buildings (Bat Conservation Trust/BMT Cordah Limited, 2005) including coach houses, stable blocks and barns (Duvergé and Jones, 2003). This species prefers to fly directly into the roost and to their roosting position and bats hang freely (Ransome and Hutson, 2000). Maternity sites are often found in large spaces at least 3–4m high, providing a sufficiently large flight area (Bat Conservation Trust/BMT Cordah Limited, 2005). This species generally uses night roosts to rest whilst foraging, which are found in a variety of structures, for example outbuildings, garages, stables, milking sheds, porches and trees (Duvergé and Jones, 1994; Ransome and Hutson, 2000; Duvergé and Jones, 2003). Male bats remain solitary through the summer and often use underground sites (Bat Conservation Trust/BMT Cordah Limited, 2005). In winter, both male and female bats choose underground sites for hibernation, including tunnels, mines, caves or cold building basements (Bat Conservation Trust/BMT Cordah Limited, 2005). Requires a range of conditions in a series of suitable hibernacula (Harris and Yalden, 2008). The main hibernation site is usually within 15km of the maternity roost, but some bats may travel up to 60km between such sites (Ransome and Hutson, 2000). Faithful to traditional summer and winter roosts (English Nature, 2003).
Lesser horseshoe bat	*Rhinolophus hipposideros*	Roost sites include attics, chimneys and boiler rooms of buildings, rural houses and outbuildings in the summer, and cellars, tunnels, disused mines and caves for hibernation (Schofield *et al.*, 2002). Also found in industrial buildings. This species prefers to fly directly into roost sites and into their roosting position (Bat Conservation Trust/BMT Cordah Limited, 2005). Maternity sites are often found in large roof spaces at least 3–4m high providing a large flight area (Bat Conservation Trust/BMT Cordah Limited, 2005). A range of conditions is required throughout the year but this may be found in one building with, for example, an attic for the summer and a cellar for the winter. Summer and winter roost sites are generally no more than 5–10km apart (Bat Conservation Trust/BMT Cordah Limited, 2005). The lesser horseshoe bat also uses alternative roost sites during the night and day.
Daubenton's bat	*Myotis daubentonii*	Roosts are found in hollow trees, bridges or sometimes buildings (Billington and Norman, 1997) generally close to water (Racey *et al.*, 1998). Nursery roosts are not exclusively female (Angell *et al.*, 2013) – males may make up 25% or more of the colony and large male-only colonies have also been recorded. Boonman (2000) found that this species selected oaks over beech trees and preferred roosts on the edges of woodlands in a study in the Netherlands. Hibernation sites are usually underground including caves, mines and suitable tunnels where bats are found both in crevices and on open walls (Altringham, 2003). They may also hibernate in tree cavities (Bat Conservation Trust/BMT Cordah Limited, 2005).

Species common name	Species scientific name	Roosting preferences
Brandt's/whiskered bat[22]	*Myotis brandtii/mystacinus*	Both species can roost in trees and a wide range of buildings in the summer (Bat Conservation Trust/BMT Cordah Limited, 2005). Hibernates in caves or other underground sites, where they can be found in the open or in cracks and crevices (Altringham, 2003).
Natterer's bat	*Myotis nattereri*	Roost sites include tree holes and different types of buildings but has also been found in bridges (Billington and Norman, 1997; Smith and Racey, 2002). Usually roost in attics between late May and mid-July (Smith and Racey, 2002) and often roosts have enough space for internal flight (Swift, 1997). This species also breeds in bat boxes (Park *et al.*, 1998; Bilston, 2014). Timber-framed barns built between the 12th and 19th centuries may be particularly important to this species (Briggs, 1995, 2002), with roosts found in mortise joints in both the summer and winter. Hibernates in cracks and crevices in caves and mines (Altringham 2003). Other hibernation sites recorded are canal and railway tunnels, ice houses and tree cavities (Smith and Racey, 2002).
Bechstein's bat	*Myotis bechsteinii*	Maternity roosts are found in tree holes in the canopy, generally in old trees with dead branches (Altringham, 2003). May be found in woodpecker holes in old oaks (Bat Conservation Trust/BMT Cordah Limited, 2005). Recorded switching roosts frequently (Kerth *et al.*, 2001; Reckardt and Kerth, 2007)). One study recorded roosts in rot holes, woodpecker holes and in a gap behind thick ivy (Palmer, 2013). A study of ten colonies across the Isle of Wight found 90% of maternity roosts in woodpecker holes in ash trees (Davidson-Watts, 2008). Another study found a maternity roost in a woodpecker hole in an oak tree on a golf course (Davidson-Watts, 2014b). Hibernates in trees and sometimes caves or other underground sites (Harris and Yalden, 2008). Chilmark Quarry is an example of Bechstein's bats using an abandoned mine for hibernation.[23]
Noctule	*Nyctalus noctula*	Roosts almost exclusively in tree holes, but sometimes found in bat boxes or buildings (Altringham, 2003). One Netherlands study found that woodpecker holes are preferred, in trees close to woodland edge (Boonman, 2000). Hibernates in trees but sometimes found in buildings (Bat Conservation Trust/BMT Cordah Limited, 2005).
Leisler's bat	*Nyctalus leisleri*	Roosts in trees, bat boxes and buildings such as houses; for example, around the gable end of lofts, under tiles, under soffit boards and in disused chimneys (Corbet and Harris, 1991). Often uses a variety of sites in the summer (Waters *et al.*, 1999). Hibernates in tree holes, buildings and sometimes underground sites (Bat Conservation Trust/BMT Cordah Limited, 2005).

[22] Brandt's and whiskered bats were only separated in 1971. Their ecologies are apparently similar although further research is needed.
[23] http://jncc.defra.gov.uk/protectedsites/sacselection/sac.asp?EUCode=UK0016373

Species common name	Species scientific name	Roosting preferences
Common pipistrelle and soprano pipistrelle	*Pipistrellus pipistrellus and P. pygmaeus*	Maternity colonies are found mainly in buildings, usually roosting out of sight in crevices. Colonies may use a number of sites through the summer but are often loyal to the same sites for many years (Thompson, 1992). Maternity colonies are extremely variable in terms of numbers, from 20 to over 1,000 bats (Speakman *et al.*, 1999). Barlow and Jones (1999) found that soprano pipistrelle colonies (median of 203) tended to be larger than those of the common pipistrelle (median of 76). Davidson-Watts *et al.* (2006) reported common pipistrelle shifting roosts between pregnancy and lactation. Davidson-Watts (2007) found that roost selection was based on temperature for common pipistrelle and on surrounding habitats (woodland and water) for both species. Males roost singly or in small groups in the summer, in buildings or trees (Lundberg and Gerell, 1986). Bat boxes are used by both males and females but generally only males use them during the summer (Park *et al.*, 1998). These species do not use underground sites for hibernation but are sometimes found in the cracks and crevices of buildings in the winter (Bat Conservation Trust/BMT Cordah Limited, 2005). Evidence from the Netherlands shows mass swarming events of common pipistrelle bats in the autumn followed by mass hibernation in a diverse range of building types in urban environments (Korsten *et al.*, 2015). This phenomenon requires some research in the UK but ecologists should be aware of the potential for larger numbers of this species to be present during the autumn and winter in large buildings in highly urbanised environments.
Nathusius' pipistrelle	*Pipistrellus nathusii*	The very few known British nursery roosts are in buildings, with hibernation roosts in hollow trees and crevices in cliffs, walls and caves (Altringham, 2003). One study recorded males roosting under lead flashing and roof tiles (Hargreaves, 2012).
Serotine	*Eptesicus serotinus*	Roosts in buildings in small cavities or crevices with high access points such as gables but occasionally also found in trees (Bat Conservation Trust/BMT Cordah Limited, 2005). Recorded hibernation sites include cavity walls, disused chimneys and occasionally caves (Bat Conservation Trust/BMT Cordah Limited, 2005).
Barbastelle	*Barbastella barbastellus*	In summer, breeding females move regularly (Greenaway, 2008) between a large number of different tree roosts (Billington, 2003; Greenaway, 2001; Zeale, 2011). One study found that they preferred dead trees surrounded by holly understorey (Greenaway, 2001) and another found them in tree crevices and cavities, between overlapping limbs and behind ivy, on average 6.9m above ground level (Billington, 2003). Greenaway (2008) found that tree roosts were in relatively undisturbed places and frequently in thick cover, although cracks much higher up in trees were used at the time of birth. Bat boxes are also used (Greenaway, 2008). Davidson-Watts (2008, 2014a) reported almost all roosts found in two studies were behind loose bark and in mixed locations not always surrounded by understorey. Winter roosts include deep, hollow trees (usually dead and among holly understorey) and sometimes buildings or underground sites (Greenaway, 2001). Other winter roosts recorded are flaking bark and splits less than 2m above the ground (Billington, 2000) and disused railway tunnels, barns, outbuildings, church porches and lime kilns. Chilmark Quarry is an example of barbastelle bats using an abandoned mine for hibernation.[24] Spring and autumn roosts have been recorded behind loose bark (Billington, 2000; Greenaway, 2001), in dead tree stumps (Greenaway, 2001) and in splits in limbs mainly less than 2m above ground level (Billington, 2000).

[24] http://jncc.defra.gov.uk/protectedsites/sacselection/sac.asp?EUCode=UK0016373

Species common name	Species scientific name	Roosting preferences
Brown long-eared bat	*Plecotus auritus*	Maternity roosts found in trees, in the voids of large, old buildings and bat boxes in woodland (Briggs, 1995; Bilston, 2014). Usually roosts against wooden beams at the roof apex in attics or farm buildings (Bat Conservation Trust/BMT Cordah Limited, 2005). Bats often cluster at the highest part of the roof and require enough space for unobstructed, internal flight (Entwistle *et al.*, 1997). Shows high roost fidelity (Entwistle *et al.*, 1997). Commonly uses feeding perches and night roosts in porches or outbuildings separate from the main roost (Bat Conservation Trust/BMT Cordah Limited, 2005). Hibernate in underground sites, tree holes and buildings (Bat Conservation Trust/BMT Cordah Limited, 2005).
Grey long-eared bat	*Plecotus austriacus*	Frequently roosts on ridge beam in spaces between rafters. Maternity colonies show high roost fidelity (Razgour *et al.*, 2013). Number of males in maternity colony increases through summer. Many males are, however, solitary.

3.5 Species emergence times

Table 3.3 provides information on the emergence times of different bat species. The time of emergence from a roost depends on the species' ecology, the amount of protective cover around the roost, the reproductive status of the bats in question, and the ambient weather conditions on the night in question and on previous nights. For some species, there is a fine balance between the need to forage and vulnerability to predators, and therefore absolute times after sunset have not been provided. Instead, the species have been separated into 'earlier' and 'later' emerging species. It should be noted that species known to exit roosts later may actually exit the roost itself earlier but remain under cover until it gets dark. The behaviour where bats appear to fly back and forth to 'test' light levels before fully emerging is often termed 'light sampling', but its actual function is unknown.

Table 3.3 Approximate emergence times of different UK species.

Earlier emerging species	Later emerging species
Noctule	Lesser horseshoe bat
Leisler's bat	Greater horseshoe bat
Common pipistrelle	Daubenton's bat
Soprano pipistrelle	Brandt's bat
Nathusius' pipistrelle	Whiskered bat
Serotine	Natterer's bat
	Bechstein's bat
	Barbastelle
	Brown long-eared bat
	Grey long-eared bat

3.6 Species foraging habitat preferences

Table 3.4 provides information on the foraging habitat preferences of different bat species. As foraging is influenced by the availability and quality of habitat around the roost, the time of year (linked to seasonal prey abundance) and the ambient conditions on the night in question this table should not be considered exhaustive (and was not derived from a thorough literature search). Bats have also been found in open landscapes such as farmland, mires, moorlands and coastal cliffs.

Table 3.4 Foraging habitat preferences and foraging strategies of different UK species.

Species	Foraging habitat preferences (with commuting preferences added for some species)
Lesser horseshoe	Preferred foraging habitats include broadleaved woodland well connected by commuting routes such as hedges, woodland edge and riparian trees (Bontadina *et al.*, 2002; Schofield *et al.*, 2002). This species has also been recorded in coniferous woodland (Schofield *et al.*, 2002). Probably reluctant to cross open space as recorded flying very low (less than 1m) in open habitats (Schofield *et al.*, 2002). This species can remain active during the hibernation period (Williams, 2001).
Greater horseshoe	Preferred foraging habitats are ancient semi-natural and deciduous woodland and cattle-grazed pastures (Duvergé and Jones, 1994; Ransome, 1997; Duvergé and Jones, 2003). Bats tend to forage on the boundaries of grazed pastures and woodland, tree lines or tall, thick hedgerows (Ransome, 1997). One study showed that bats fly close to field boundaries and reduce their flight height when out in the open (Duvergé and Jones, 2003). A spring study showed grazed pastures and broadleaved woodland were selected over other habitats (Flanders and Jones, 2009). This species can remain active during the hibernation period (Park *et al.*, 1999).
Daubenton's bat	Preferred foraging habitat is over water (Jones and Raynor, 1988): this species favours riverine habitats (Racey and Swift, 1985; Rydell *et al.*, 1994) but is also known to forage in woodland.
Whiskered / Brandt's bat	Buckley *et al.* (2013) found whiskered bat used mixed woodland, riparian vegetation, arable and rough grassland habitats although selected the first two as core foraging habitats. Berge (2007) found that whiskered bat selected pasture with hedgerows. A German study showed Brandt's bat favours woodland and whiskered bat favours areas near rivers and more open habitats with hedges and coppices (Taake, 1984).
Natterer's bat	Preferred foraging habitat is semi-natural broadleaved woodland, tree-lined river corridors and ponds, but also uses grassland (Smith and Racey, 2002, 2008). Avoids dense coniferous plantation (Smith and Racey, 2008). An autumn study revealed the species to use woodland and mixed agricultural areas (Parsons and Jones, 2003).
Bechstein's bat	Predominantly associated with ancient broadleaved woodlands (Greenaway and Hill, 2004), with a strong association with oak and ash (Hill and Greenaway, 2005). Various studies have recorded foraging under a closed canopy (Fitzsimons *et al.*, 2002, Harris and Yalden, 2008). One recent study recorded foraging in mixed-age coppice, high forest with little understorey, unimproved grassland, a dry stream corridor with scrub and trees and tree lines and hedgerows in a pastoral landscape (Palmer, 2013). Commuting was recorded along the stream corridor and hedgerows (Palmer, 2013). Davidson-Watts (2014b) also reported use of hedgerows in grazed pasture for commuting and patches of coniferous woodland used for commuting when these were present as part of a larger broadleaved block. Davidson-Watts (2013) also reported use of tree-lined river margins.
Noctule	Found in a range of habitats, forages out in the open, often over trees and with a strong affinity to water (Altringham, 2003). Reported as selecting broadleaved woodland and pasture (Mackie and Racey, 2007).
Leisler's bat	Recorded foraging in woodland edge, scrub or woodland-lined roads and over pasture (Waters *et al.*, 1999). Also recorded over drainage canals, lakes and coniferous forests (Shiel *et al.*, 1999). Recorded as selecting parkland/amenity grassland, deciduous woodland edge and rivers/canals but avoiding improved grassland (Russ and Montgomery, 2002). One road-based study showed this species to be equally active in all habitats available (hedges, tree lines, woodland, grassland, streetlights and arable areas) (Russ *et al.*, 2003).
Common pipistrelle	Shows a preference for deciduous woodland but a generalist using a wide range of habitats (Davidson-Watts and Jones, 2006; Davidson-Watts *et al.*, 2006).
Soprano pipistrelle	Tends to select riparian habitats over other habitat types available (Davidson-Watts and Jones, 2006; Davidson-Watts *et al.*, 2006).

Species	Foraging habitat preferences (with commuting preferences added for some species)
Nathusius' pipistrelle	Riparian habitats, broadleaved and mixed woodland and parkland, occasionally found in farmland but always near water (Harris and Yalden, 2008). Found over lakes and rivers (Vaughan *et al.*, 1997). One study recorded males feeding over lake edge and managed gardens and fields around a lake (Hargreaves, 2012).
Serotine	Catto *et al.* (1996) and Robinson and Stebbings (1997) identified the following habitats as important for foraging: cattle pasture, playing fields, village greens, white streetlights, tree-lined hedgerows and woodland edge.
Barbastelle	Forages over/in riparian zones, broadleaved woodland, unimproved grassland and field margins (Zeale, 2011; Zeale *et al.*, 2012). Foraging has also been recorded at an irrigation reservoir, ponds in woodland, areas of set-aside, floodplain habitats, a sewage farm and a pumping station (Greenaway, 2008). Bats tend to wait for darkness to cross open areas (Greenaway, 2008). However, barbastelle avoided wetlands, preferring woodlands and treelines in a study by Davidson-Watts (2014a).
Brown long-eared	Strongly associated with tree cover (Entwistle *et al.*, 1996), prefers woodland with cluttered understorey including native species, particularly deciduous (Murphy *et al.*, 2012). Also forages in mixed woodland edge and among conifers. Use of hedgerows increases through the active season (Murphy *et al.*, 2012).
Grey long-eared	Prefers to forage in more open or edge habitats, including unimproved lowland grassland (meadows and marshes), wooded riparian vegetation and broadleaved woodland (woodland mainly used in low temperatures or heavy rainfall) (Razgour *et al.*, 2011, 2013). In agricultural habitats, forages along field margins, hedgerows and scattered trees.

3.7 Species Core Sustenance Zones

BCT has been working on defining Core Sustenance Zones (CSZs) for different bat species through an extensive literature review (see Table 3.5). A CSZ refers to the area surrounding a communal bat roost within which habitat availability and quality will have a significant influence on the resilience and conservation status of the colony using the roost.

With reference to development, the CSZ could be used to indicate:

- The area surrounding a communal roost within which development work may impact the commuting and foraging habitat of bats using that roost.
- The area within which it may be necessary to ensure no net reduction in the quality and availability of foraging habitat for the colony.

Consideration should be given to the extent of a background data search in relation to the species likely to be present and the impact of the development (see Section 4.2.2). CSZs could also be used to interpret the results of background data searches (see Section 4.2.3).

More information on how these CSZs have been derived can be found on the BCT website.[25]

Table 3.5 CSZs for different UK bat species.

Species	CSZ radius (km)	No. of bats studied	No. of studies	Confidence in zone size[a]
Lesser horseshoe[b]	2	83	4	Good
Greater horseshoe[b]	3	39	4	Moderate
Daubenton's bat	2	7	2	Poor
Whiskered/Brandt's bat	1	24	1	Poor
Natterer's bat	4	53	2	Good
Bechstein's bat[b]	1	70	4	Moderate
Noctule	4	20	1	Poor
Leisler's bat	3	20	2	Moderate
Common pipistrelle	2	23	1	Poor
Soprano pipistrelle	3	91	3	Good
Nathusius' pipistrelle	3	9	2	Poor
Serotine	4	13	1	Poor
Barbastelle[b]	6	69	3	Moderate
Brown long-eared	3	38	1	Poor
Grey long-eared[b]	3	20	1	Moderate

[a] Confidence is based on the number of bats and number of studies used to inform the calculation of CSZ.

[b] There may be justification with Annex II and other rare species to increase the CSZ to reflect use of the landscape by all bats in a population. We would suggest increasing the CSZ of Bechstein's bat to at least 3km, reflecting its specific habitat requirements.

[25] http://www.bats.org.uk

3.8 Species population estimates, distribution and status

Data collected on the presence and abundance of bat species should be assessed in the context of any available knowledge about the distribution and rarity of local, county and national bat populations. Without this context it is not possible to make an assessment about the conservation significance of the survey findings. Potential sources of data on distribution and rarity of bat species are given in Table 3.6.

Table 3.6 Potential sources of data on species distribution and bat population status at different geographic scales.

Geographic scale	Sources of data on species distribution and bat population status at relevant scale
Local	Background data search (see Chapter 4 for different sources of data)
	Local Biodiversity Action Plans
	Local Mammal Atlas
	Data from ecological reports submitted with planning applications
	Local Records Centre
County	County Bat Group
	County Wildlife Trust
	County Recorder
	Local Records Centre
Country	Article 17 Reporting[26]
UK / Great Britain	Article 17 Reporting
	National Bat Monitoring Programme
	Richardson, 2000
	Harris and Yalden, 2008
	Dietz *et al.*, 2011
	Wray *et al.*, 2010

3.9 Species-specific considerations

A few bat species are difficult to detect with bat detectors because they produce quiet (low amplitude) echolocation calls, have very directional echolocation calls, or sometimes use their eyes or ears rather than echolocation (especially in or close to roosts or when gleaning prey). Longer sampling periods, including the use of automated/static detectors, will increase the likelihood of detecting these species acoustically. Other methods include DNA analysis of droppings (where possible) or advanced bat licence survey techniques (see Chapter 9). Table 3.7 provides information on echolocation call characteristics for species with low-amplitude calls and suggests solutions to overcome this limitation.

Table 3.7 Bat species that are difficult to detect with bat detectors and methods to overcome this limitation.

Species	Echolocation call characteristics which create low likelihood of detection	Potential solutions to this limitation
Lesser horseshoe bat[a]	Calls are directional at high frequency and are subject to a marked degree of attenuation that reduces potential detection distance and the likelihood of a bat being detected if echolocation calls are received by the microphone significantly off-axis.	Full-spectrum recording is recommended. Deploying an automated/static detector within constrained flight corridors such as tunnels and natural corridors through vegetation that are often used by this species and where flights are concentrated will increase the likelihood of recording bats.

[26] Member states of the European Union are required to report on the implementation of the Habitats Directive every six years through what is known as Article 17 reporting. Article 17 reports are available for the UK and for England, Wales, Scotland and Northern Ireland separately and include data on population estimates, range, distribution and status of the different bat species, with information taken from a number of sources. The latest reporting at the time of writing was JNCC, 2013 (reporting on the period 2007–2012) and the relevant reports can be found on the JNCC website (http://jncc.defra.gov.uk/page-6387).

Species	Echolocation call characteristics which create low likelihood of detection	Potential solutions to this limitation
Bechstein's bat	Calls of *Myotis* species for which call intensity has been measured are of fairly low amplitude (Faure *et al.*, 1990) and are generally frequency modulated (FM – where energy is spread across multiple frequencies). When in woodland this species is likely to spend a proportion of its time high in the tree canopy, making it potentially difficult to detect.	Even if its calls can be recorded, separating Bechstein's bat from other *Myotis* species is difficult (or impossible) by acoustic analysis (Parsons and Jones, 2000; Walters *et al.*, 2012). Catching surveys, aided by an acoustic lure, are likely to be required where there is a reasonable potential for this species to be present (i.e. habitat is suitable and a site is within the known geographic range) if this species may be at risk from a proposal.
Barbastelle	Very low-intensity echolocation calls (Goerlitz *et al.*, 2010). Flight is relatively fast, so recordings tend to be of short duration.	Use of broad-band recordable detectors has helped to demonstrate that this species is present more frequently and across a wider range of habitats than previously believed. Calls are often missed by ecologists listening in the field as they are often indistinct, not repeated and masked by calls of other species. It is essential to use recordable detectors with this species. Attempt to intersect bats with detectors on commuting routes, when calls are potentially of higher intensity.
Brown or grey long-eared bat	Low-amplitude and FM calls are often used. Foraging bats often make no sound and use eyes or ears to hunt by gleaning (Swift and Racey, 2002). Additionally, difficult to detect whilst foraging in understorey.	Attempt to intercept bats with detectors on commuting routes, when calls are potentially of higher intensity. Night vision or infrared camera equipment can be used to identify long-eared species bats by their distinctive appearance. Inside buildings, placing a detector high up usually increases the number of passes recorded.

[a] Similar issues for greater horseshoe bat but reduced due to lower-frequency calls than for lesser horseshoe bats.

Recent research by Scott and Altringham (2014) analysed the probability of detection of different species according to the intensity and directionality of their calls in woodland habitats. Table 3.8 provides information on the number of surveys required to achieve 95% certainty of detection of different species on walked transect surveys in the study (in woodland habitats, using Pettersson D500x and D240x detectors and software developed for the project to automatically isolate and identify bat calls). This table is included to illustrate the relative likelihood of picking up different species rather than to recommend the protocol, which was developed specifically for monitoring purposes.

Table 3.8 Number of surveys required to achieve 95% certainty of detection on walked transect surveys in woodland (Scott and Altringham, 2014).

Species	Number of surveys to achieve 95% certainty of detection for walked transect survey
Pipistrelle	1
Brandt's bat	2
Whiskered bat	2
Barbastelle	2
Horseshoe bat	4
Natterer's bat	5
Brown long-eared bat	Up to 9[a]
Bechstein's bat	4–6[b]
Alcathoe	2–3[b]

[a] It may be reasonable to assume that brown long-eared bats are likely to be present in most broadleaved woodland. Alternative methods (such as existing records or trapping surveys) may be more effective if proof of presence is required.
[b] Untested estimate.

Chapter 4

Preliminary ecological appraisal for bats

4.1 Introduction

A project often starts with a preliminary ecological appraisal covering ecological features of interest (although smaller projects may not require all elements of a preliminary ecological appraisal, as discussed below). CIEEM has published *Guidelines for Preliminary Ecological Appraisal* (CIEEM, 2013d). These guidelines acknowledge that there is a wide range of terminology used for such surveys but that their purpose is to:

- establish baseline conditions and determine the importance of ecological features present within the specified area;
- establish any requirements for detailed/further surveys (e.g. for bats);
- identify key project constraints to make recommendations for design options to minimise impacts; and
- identify mitigation measures (as far as possible) and enhancement opportunities.

Preliminary ecological appraisals generally include a desk study and fieldwork, often based on the Phase I survey method (JNCC, 2010). The preliminary ecological appraisal is generally extended to identify habitats present that have the potential to support protected species.

As with all surveys, survey design should be based around the questions that require answers. The main questions with respect to preliminary ecological appraisal for bats generally relate to assessing what the potential impacts of the proposal are on bats both on and off site and include the following:

- Is the site close to any internationally or nationally designated sites for bats or with bats as part of the reason for designation?
- Which species are known from the area, what is their conservation status and what types of habitats are they likely to be found in?
- Are there likely to be species listed in Annex II of the Habitats Directive?
- Are there likely to be species particularly at risk of being impacted by the type of activities proposed?
- What habitat types are present on site and in the surrounding area that are (a) likely to be used by bats for roosting, foraging or commuting, and (b) likely to be impacted by the proposal?
- What is the likely suitability of those habitats for bats?
- How do the habitats on site connect to habitats in the surrounding area to create an ecological network?

In order to answer the questions outlined above, a preliminary ecological appraisal for bats, consisting of a desk study and fieldwork, is generally carried out. This is described in the following sections. This assessment will enable an ecologist to proceed with further bat surveys as necessary using an iterative approach where each stage informs the next.

A full preliminary ecological appraisal for bats may not be necessary for smaller projects (e.g. projects impacting a single house or barn). Relevant elements, such as a study of maps, aerial photographs and site photographs, may provide enough information to skip straight to a preliminary roost assessment (see Chapters 5 and 6) without a preliminary ecological appraisal and with elements of the desk study (such as a background data search from a Local Records Centre (LRC)) carried out afterwards if potential for bats or evidence of bats is found. This is likely to save both time and financial resources.

4.2 Preliminary ecological appraisal – desk study

4.2.1 Sources of information for desk study

The aim of a desk study for bats is to collate and review existing information about a site and its surroundings to inform the design of subsequent bat surveys and inform the impact assessment for the project.

When using or referring to materials obtained from external sources, rules of copyright should be noted and adhered to. There may also be restrictions on the commercial use of Internet resources.

This information includes the following:

- Photographs and descriptions of the site.
- Maps and aerial photographs can be viewed using applications such as Google Maps[27] or Bing Maps,[28] both of which also provide a street view option. These allow an ecologist to identify habitats and features that are likely to be important for bats and assess their connectivity. Note when the photographs were taken; if old, conditions may have changed.
- Records of statutory and non-statutory designated sites (where bats form all or part of the reason for the designation) can be found on the Multi Agency Geographic Information for the Countryside (MAGIC) website,[29] although less information is provided for Scotland. Scottish users should refer to the SNH site link system.[30]

[27] https://www.google.co.uk/maps/
[28] https://www.bing.com/maps/
[29] https://www.magic.defra.gov.uk
[30] https://gateway.snh.gov.uk/sitelink/

It is usually necessary to contact the LRC or LPA to obtain records of non-statutory sites such as County Wildlife Sites or Sites of Importance for Nature Conservation; these are often designated for botanical reasons but their descriptions can provide useful information about habitats and may contain records of bats. LRCs are found in most counties and generally charge a fee to search for records of designated sites and protected species. A list of active centres can be found on the website of the National Forum for Biological Recording (NFBR).[31]

- Records of bats in the area can be obtained from a number of organisations by providing a grid reference or site boundary and stating the required radius for the search and the type of records required. It is important to note that the absence of bat records does not confirm the actual absence of bats because records are not always collected in a systematic and thorough way. Organisations that hold local bat records are listed below.
  - National Biodiversity Network (NBN).[32] The use of NBN data in commercial ecological reports is not permitted under the NBN code of conduct.[33]
  - LRCs (see above).
  - LBGs – found in most counties, sometimes have a database of records or a county bat distribution atlas, will sometimes carry out a background data search for a fee although many share their records with LRCs, may also provide information on the local and regional status of populations; contact details for each LBG can be obtained from the BCT website[34] (search for 'local bat groups').

  Other sources of bat records or information may include the following:
  - County Ecologists (or Biodiversity or Nature Conservation Officers) – employed by some local, county or district councils.
  - Local Wildlife Trusts (LWTs).[35]
  - County mammal recorders – volunteers who collate records of mammal sightings in their county; contact details are available from the Mammal Society website.[36]
  - Local publicly funded research projects, e.g. data from all Natural Environment Research Council funded research projects on bats are published/available free of charge online.
  - Other planning applications for the area – may provide some insight into local bat species and activity levels; planning applications can be found on county/district/borough council websites.
  - The MAGIC website[37] now provides information on EPS licences.
  - Local or national mining history or caving groups and clubs, and caving councils – these may have useful information on hibernation roosts and some cave systems have biological recorders who publish records in club or regional journals; see the BCA's website[38] for information.
  - On-site personnel such as site security guards, caretakers or gardeners – may provide anecdotal evidence that gives useful pointers, although data may not be reliable enough to be used in a preliminary ecological appraisal.
- Other relevant literature – for example, species distribution and status (see Section 3.8). This information is particularly important when analysing survey data and carrying out an impact assessment.

4.2.2 Geographical extent of desk study

As a minimum, it is recommended that background data searches should be carried out up to 2km from the proposed development boundary (including all temporary works). However, the data search should be related to the scheme's ZoI (see Section 2.2.3) and consider the CSZs of species likely to be present (see Section 3.7), and may need to extend up to 10km for larger projects.

Statutory designated sites such as SACs or SSSIs relevant to bats within 10km should also be considered.

Some other considerations that should be applied to background data searches are as follows:

- In areas where bat roosts and foraging areas are more sparsely distributed, the background data search radius may need to be increased.
- In coastal areas, migrating bats may need to be considered. Ringing has now confirmed that some of our bat species migrate between the UK and the continent.[39]

4.2.3 Interpretation of desk study data

The desk study records provide contextual information for the survey design stage as well as the evaluation of the survey results. They should be interpreted to identify:

- if proposed activities are likely to impact on a SAC or the qualifying feature of a SAC (this may trigger the need for a HRA);
- if the proposed activities are likely to impact on other designated sites and thus require consultation with relevant bodies;[40]
- any species (or genera) confirmed/thought to be present;
- any bat roosts that will be impacted (on or off-site);
- if it is likely that the CSZs of bats from roosts off-site will be impacted (see Section 3.7);
- if there are any rare species in the area that may require species-specific survey methodologies.

4.2.4 Next steps

It is usual for a desk study to be followed by the fieldwork element of a preliminary ecological appraisal (although, as discussed above, this may not be needed for smaller projects).

There may be some cases where aerial photographs and descriptions of the site confirm there is no habitat suitable for bats on site or in

31 https://www.nfbr.org.uk
32 https://www.nbn.org.uk
33 https://data.nbn.org.uk/Terms
34 https://www.bats.org.uk
35 https://www.wildlifetrusts.org
36 https://www.mammal.org.uk
37 https://www.magic.defra.gov.uk
38 https://www.british-caving.org.uk
39 A Nathusius' pipistrelle that was ringed near Bristol in the UK in 2012 was subsequently found in the Netherlands, 600km away (Hargreaves, 2014). Other evidence is emerging that supports the theory that bat species migrate between the UK and the continent (BSG Ecology, 2013a, 2014a, b).
40 NE has developed the concept of Impact Risk Zones (IRZs) around SSSIs. They define zones around each SSSI (found here: http://magic.defra.gov.uk/MagicMap.aspx) which reflect the particular sensitivities of the features for which it is notified and indicate the types of development proposal which could potentially have adverse impacts. The IRZs also cover the interest features and sensitivities of European sites. More information on IRZs can be found here: https://www.gov.uk/construction-near-protected-areas-and-wildlife

the surrounding area. Ecologists and their clients may want to keep a record of the rationale behind the decision not to survey.

4.3 Preliminary ecological appraisal – fieldwork

4.3.1 Description and aims

A preliminary ecological appraisal for bats is a walkover of the proposed development site to observe, assess and record any habitats suitable for bats to roost, commute and forage both on site and in the surrounding area (it is important that connectivity within the landscape is also considered at this stage). The aim is to determine the suitability of a site for bats, to assess whether further bat surveys will be needed and how those surveys should safely be carried out.

4.3.2 Equipment

Generic documentation/equipment required for field surveys for bats is provided in Section 2.5.2; survey-specific equipment is listed in Appendix 1.

4.3.3 Expertise and licences

Section 2.5.1 discusses expertise and Section 1.2.2 provides information on licences. Unless an ecologist intends to enter buildings or investigate PRFs in trees with a torch or endoscope, a preliminary ecological appraisal is unlikely to cause disturbance so a licence is generally not needed.

4.3.4 Methods

Ecologists should identify and record any structures and trees that could be suitable for bats to roost in and any habitats that could be suitable for bats to commute, forage or swarm in/at. If suitability is assessed at this stage, the scheme presented in Table 4.1 should be used. Please note that low suitability roosting habitats may be present in commuting/foraging habitats that are of high suitability, and vice versa. Roosting habitats and commuting/foraging habitats should be assessed separately and independently.

Table 4.1 Guidelines for assessing the potential suitability of proposed development sites for bats, based on the presence of habitat features within the landscape, to be applied using professional judgement.

Suitability	Description Roosting habitats	Commuting and foraging habitats
Negligible	Negligible habitat features on site likely to be used by roosting bats.	Negligible habitat features on site likely to be used by commuting or foraging bats.
Low	A structure with one or more potential roost sites that could be used by individual bats opportunistically. However, these potential roost sites do not provide enough space, shelter, protection, appropriate conditions[a] and/or suitable surrounding habitat to be used on a regular basis or by larger numbers of bats (i.e. unlikely to be suitable for maternity or hibernation[b]). A tree of sufficient size and age to contain PRFs but with none seen from the ground or features seen with only very limited roosting potential.[c]	Habitat that could be used by small numbers of commuting bats such as a gappy hedgerow or unvegetated stream, but isolated, i.e. not very well connected to the surrounding landscape by other habitat. Suitable, but isolated habitat that could be used by small numbers of foraging bats such as a lone tree (not in a parkland situation) or a patch of scrub.
Moderate	A structure or tree with one or more potential roost sites that could be used by bats due to their size, shelter, protection, conditions[a] and surrounding habitat but unlikely to support a roost of high conservation status (with respect to roost type only – the assessments in this table are made irrespective of species conservation status, which is established after presence is confirmed).	Continuous habitat connected to the wider landscape that could be used by bats for commuting such as lines of trees and scrub or linked back gardens. Habitat that is connected to the wider landscape that could be used by bats for foraging such as trees, scrub, grassland or water.
High	A structure or tree with one or more potential roost sites that are obviously suitable for use by larger numbers of bats on a more regular basis and potentially for longer periods of time due to their size, shelter, protection, conditions[a] and surrounding habitat.	Continuous, high-quality habitat that is well connected to the wider landscape that is likely to be used regularly by commuting bats such as river valleys, streams, hedgerows, lines of trees and woodland edge. High-quality habitat that is well connected to the wider landscape that is likely to be used regularly by foraging bats such as broadleaved woodland, tree-lined watercourses and grazed parkland. Site is close to and connected to known roosts.

[a] For example, in terms of temperature, humidity, height above ground level, light levels or levels of disturbance.
[b] Evidence from the Netherlands shows mass swarming events of common pipistrelle bats in the autumn followed by mass hibernation in a diverse range of building types in urban environments (Korsten *et al.*, 2015). This phenomenon requires some research in the UK but ecologists should be aware of the potential for larger numbers of this species to be present during the autumn and winter in large buildings in highly urbanised environments.
[c] This system of categorisation aligns with BS 8596:2015 Surveying for bats in trees and woodland (BSI, 2015).

Assessment of suitability, carried out as per the table above, informs the design of subsequent survey work, although the elements outlined in Section 2.2 should also be considered, in particular the potential impacts (Section 2.2.2) and proportionality (Section 2.2.5). The assessment of suitability will be further refined for roosts during a preliminary roost assessment (Sections 5.2 and 6.2). These assessments inform subsequent survey effort for roosts (see Sections 7.1.7 and 7.1.8) and commuting and foraging habitats (see Section 8.2.7). The early assessment of suitability for bats, however, should not be confused with the later assessment of the conservation value of a site, which relates to the species, numbers and roost types **actually present**.

During the preliminary ecological appraisal, the ecologist should consider the further surveys needed (if any), their logistics (resources, emergence survey locations, transect routes, static detector locations, timings), and any potential health and safety hazards reported.

If no suitable habitat for bats is found, then further surveys are not likely to be necessary. Ecologists and their clients may want to keep a record of the rationale behind the decision not to carry out further surveys, including evidence that an adequate assessment has been made by a suitably qualified ecologist and the conclusion is reasonable.

4.3.5 Timing

A preliminary ecological appraisal survey for bats should be done during daylight; sufficient time should be allowed to walk the entire site. It may be necessary to use multiple ecologists if only a limited amount of time is available and the site is very large.

The survey can be done at any time of year but it is recommended that at least some of the results of the desk study are available to assist in planning and carrying out the survey and before making decisions about subsequent surveys.

4.3.6 Survey effort

The survey area should be determined by the ZoI and the nature of the proposals.

4.3.7 Weather conditions

The preliminary ecological appraisal can be carried out under any weather conditions, providing that the weather conditions do not affect the ecologist's ability to carry out the survey effectively and safely.

4.3.8 Next steps

The preliminary ecological appraisal informs the design of subsequent, more detailed surveys. The following questions should be considered:

- Are further, more detailed bat surveys needed?
- What types of detailed bat surveys would be appropriate to enable the impact assessment that is needed relative to the nature and current status of the project?
- Are any specialist techniques required arising from the potential presence of particular species; for example, the use of acoustic lures to detect the presence of Bechstein's bat?
- Are any specialist techniques required arising from the presence of particular habitats: for example, the need for confined spaces training to access underground sites?
- Are any specialist techniques required arising from the potential for project-specific impacts; for example, the need to survey at crossing points on a proposed road scheme or at height?

Where further surveys and mitigation are required, the preliminary ecological appraisal report in isolation will not be adequate for submission to an LPA in support of a planning application. The report will only be adequate for this purpose if there is no need for further surveys and mitigation.

Chapter 5

Bat roost inspection surveys – buildings, built structures and underground sites

5.1 Introduction

This chapter provides information on carrying out inspection surveys for bat roosts in buildings, built structures and underground sites, collectively referred to as structures.

These surveys may be required where development proposals include demolition of a structure or a structure will be modified in such a way that bats or their roosts could be *directly* impacted if present.

These surveys may also be needed where bats roosting in a structure could be *indirectly* impacted by development activities outside the roost such as lighting/removal of vegetation or the construction of a new road/railway, where collision impacts are a possibility. In these cases it is necessary to consider whether bat roosts both on and off site may be indirectly impacted and consider surveying at least for maternity and hibernation roosts and swarming sites where appropriate.

The above principles apply regardless of the size of the development.

Roost surveys of structures should be designed to answer specific questions, such as:

- Are actual or potential bat roosts present (and if so, where)?
- Which bat species use the site for roosting?
- How many bats do these roosts support?
- Where are the bat roost access points?
- Where are the bat roosts and how do the bats get to them from the access points?
- What is the current arrangement of vegetation and lighting in relation to the access points?
- At what times of the year are bats present? How does use change seasonally?
- What types of bat roost (see Section 3.3) are present?

Answering some or all of these questions allows an ecologist to carry out an impact assessment and design a mitigation, enhancement and monitoring strategy, where relevant.

Roost surveys of structures generally take a staged approach, with the first step being a **preliminary roost assessment** (possibly preceded by a preliminary ecological appraisal – see Chapter 4), which may be followed up by **winter hibernation**, **presence/absence** and/or **roost characterisation surveys**. The latter two survey types are covered in Chapter 7, which also covers trees. Survey design should be iterative; each stage informing the next, as per the flow chart provided in Figure 5.1. The effectiveness of the surveys should be considered at each stage.

Figure 5.1 Flow chart illustrating the process used to establish which types of surveys are necessary for roosts in structures.

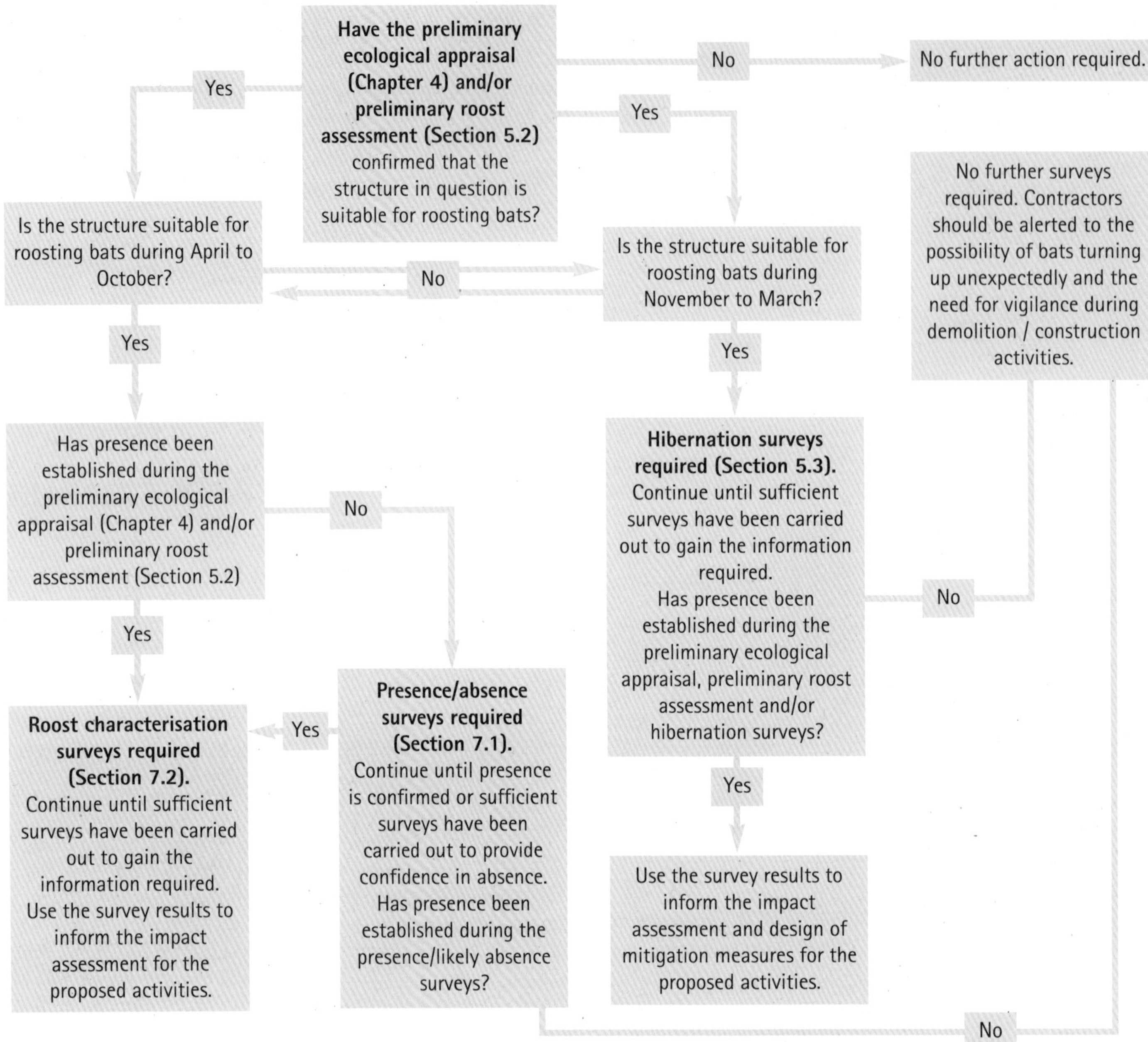

Note on Figure 5.1: In some situations bats may use the same structure throughout the year and in these situations, both arms of the flow chart need to be fully considered.

5.2 Preliminary roost assessment – structures

5.2.1 Description and aims

A **preliminary roost assessment** is a detailed inspection of the exterior and interior of a structure to look for features that bats could use for entry/exit and roosting and to search for signs of bats. The aim of this survey is to determine the actual or potential presence of bats and the need for further survey and/or mitigation. In many situations it is not possible to inspect all locations where bats may be present and therefore an absence of bat evidence does not equate to evidence of bat absence.

5.2.2 Equipment

Generic documentation/equipment required for field surveys for bats is provided in Section 2.5.2; survey-specific equipment is listed in Appendix 1.

PPE for entering confined spaces, entering spaces with asbestos, working at height or working in derelict buildings may also be required but specialist advice and training should be sought in such scenarios. More on health and safety can be found in Section 2.7.

5.2.3 Expertise and licences

Section 2.5.1 discusses expertise and Section 1.2.2 provides information on licences. The fieldwork involved in a preliminary roost assessment could result in disturbance to bats and therefore it is good practice for ecologists to hold a survey licence. The use of endoscopes requires specific training and the relevant licence; in England this would be a Class Two licence (see Section 1.2.2). Bat handling should only be carried out by ecologists licensed to handle bats or their trainees and only when the information cannot be gained by any other method. Hibernating bats, heavily pregnant bats or bats with dependent young should not be handled.

Training relating to health and safety may also be required for preliminary roost assessments; examples include the safe use of ladders or asbestos awareness training (see Section 2.7).

5.2.4 Methods

The method involves a detailed external and internal inspection of the structure to compile information on potential and actual bat entry/exit points; potential and actual bat roosting locations; any evidence of bats found and the number of ecologists that will be required for any subsequent surveys. The *Bat Workers' Manual* (Mitchell-Jones and McLeish, 2004) provides useful diagrams and definitions of architectural terms.

Sufficient time should be allowed to complete the external and internal inspection during daylight hours. The inspection should be carried out systematically and consistently through all parts of the structure and the results recorded in a standard format.

Definitions of suitability of roosting habitat are provided in Table 4.1 (page 35). The evaluation at this stage is more precise than during the preliminary ecological appraisal (see Chapter 4).

5.2.4.1 External survey

A systematic search should be made of the exterior of the structure to identify potential or actual bat access points and roosting places (although it should be noted that some may not be visible from ground level) and to locate any evidence of bats such as live or dead specimens, bat droppings, urine splashes, fur-oil staining and/or squeaking noises. Bat specimens and droppings are the most reliable type of evidence; the other types are not always the result of bat activity. **Sometimes bats leave no visible sign of their presence on the outside of a building (and even when they do, wet weather can wash evidence away).**

The search should include the ground, particularly beneath potential access points, any windowsills, window panes, walls, behind peeling paintwork or lifted rendering, hanging tiles, weatherboarding, eaves, soffit boxes, fascias, lead flashing, gaps under felt (even including those of flat roofs), under tiles/slates and in existing bat boxes. Any gaps in brickwork or stonework should be identified and searched because they may allow access to cavity- or rubble-filled walls. **This list is not exhaustive – all areas should be searched thoroughly and systematically.**

The status of the structure (with respect to structural integrity) should be established prior to the visit but, during the external survey, this information should be corroborated and any new information added to the risk assessment. This assessment is essential to ensure safety when entering a structure.

5.2.4.2 Internal survey

Where safe, a systematic search should be made of the interior of the structure to identify potential or actual bat access points and roosting places and to locate evidence of bats. Bat specimens (live or dead) and droppings are the most reliable type of evidence. Other evidence found can include urine splashes, fur-oil staining, feeding remains (moth wings), squeaking noises (which can sometimes alert an ecologist to an otherwise hidden roost), bat-fly (Nycteribiid) pupal cases (Hutson, 1984) or odour. These latter types of evidence should, however, not be relied upon in isolation to confirm presence. **Sometimes bats leave no visible sign of their presence even on the inside of a building, particularly where there are hidden cracks, crevices and voids.**

Ecologists should work quietly and check structures in a systematic manner, working upwards from the entrance and checking any cellar space last. Upon entering an individual space, the places bats are most likely to be should be checked first. For example, on entering a loft space, always look up and check the ridge beam and other beams for free-hanging bats first. Following this, the space should be checked systematically for evidence of bats.

In derelict or abandoned structures, all areas should be surveyed where it is safe to do so. Before entering upper floors or attics, the ceilings below should be inspected for any damage/concealed hatches that may indicate it is unsafe to walk above. It may also be necessary to seek professional advice (e.g. from a structural engineer) as to the safety of a building before entering or proceeding to upper floors and attics.

Where buildings are in use for residential or commercial purposes, it may not be necessary to inspect all of the rooms, instead concentrating on upper floors (evidence stuck to exterior windows, walls and windowsills may be more apparent from upper rooms than from the ground-level survey), roof spaces, boiler rooms or other dark spaces or spaces not in daily use.

Within rooms in buildings, ecologists should inspect:

- the floor and surfaces of any furniture or other objects;
- behind wooden panelling;
- in lintels above doors and windows;
- behind window shutters and curtains;
- behind pictures, posters, furniture, peeling paintwork, peeling wallpaper, lifted plaster and boarded-up windows; and
- inside cupboards and in chimneys accessible from fireplaces.

Frequently used roost locations within roofs include:

- the top of gable end or dividing walls;
- the top of chimney breasts;
- ridge and hip beams and other roof beams;
- mortise and tenon joints;
- all beams (free-hanging bats);
- the junction of roof timbers, especially where ridge and hip beams meet;
- behind purlins;
- between tiles and the roof lining; and
- under flat felt roofs.

Therefore a search of a roof void should pay particular attention to the floor, water tanks, stored materials and other surfaces beneath such locations to look for evidence of bats. Searching beneath and around the edges of insulation may also uncover historical evidence of bats as listed above. Any internal access to cavity or rubble-filled walls should be noted along with the range of conditions provided by a structure.

The above lists are not exhaustive – the ecologist should use professional judgement based on experience to decide where inspection is necessary.

Turning all torches off whilst in a dark space (e.g. a roof space or dark barn) will allow ecologists to look for light spilling in, which will indicate gaps that bats may use for entry points.

Sometimes a space may have been cleaned and evidence of bats may have been removed so this needs to be taken into consideration.

If any parts of a structure cannot be surveyed due to accessibility, this, and any other limitations of the inspection, should be clearly detailed in the report.

The following sections provide information on some structure-specific considerations.

5.2.4.3 Timber-framed and stone barns

Timber-framed and stone barns may be used by bats throughout the year, and can support a range of roost types for a variety of different species. Barns are often very open and tall, making preliminary assessment and detailed surveying of potential roost sites difficult and time-consuming. They may also contain farm machinery and other materials that can impede bat surveys.

When surveying barns, the features that should be given particular attention during an inspection survey include:

- gaps between ridge tiles and ridge and roof tiles, usually where the mortar has fallen out or the tiles are broken or lifted;
- the ridge area of the roof (particularly between the ridge beam and roofing material);
- lifted lead flashing associated with roof valleys, ridges and hips, or where lead flashing replaces tiles;
- spaces between external weatherboarding/cladding and the timber frame or walls;
- gaps behind window frames, lintels and doorways including the main doors;
- tenon and mortise joints between truss beams and braces and the principal support columns;
- cracks and crevices in timbers;
- gaps between stones or bricks (especially where purlins enter the wall and by the wall plate); and
- surfaces such as the ground, ledges, windows, sills or walls, machinery or stored material within the barns (which should be searched for bat droppings and/or urine spots or stains).

Close inspection of cavities and behind timbers should be undertaken using endoscopes, torches and/or mirrors. This often requires the use of ladders to access a safe working platform. Inspection of the roof timbers and ridge beam often requires binoculars and powerful torches to illuminate the roof from the ground.

5.2.4.4 Churches

Churches, because of their age, structure and location, often support bats. Survey considerations that are specific to churches are given below.

- Bats may share the main spaces of a church with worshippers (even if there is a separate roof void), therefore the internal survey should include all areas.
- Most churches are regularly cleaned, so bat droppings may be removed. Ask the cleaning staff if they are aware of any bats, find out the cleaning schedule and do not carry out a preliminary roost assessment immediately after the church has been cleaned. Search higher areas out of the reach of cleaners for evidence of bats.
- Urine splashes can leave a permanent and obvious stain on polished wooden, stone and metal surfaces. However, stains can persist for many years and so do not always indicate recent use of the church by bats.
- Features of churches are given specific terms: use the correct technical terminology in recording and reporting. The *Bat Workers' Manual* (Mitchell-Jones and McLeish, 2004) provides useful guidance, including diagrams.
- Churches may have underground crypts that are not immediately obvious but often support bats; enquire about the existence of underground spaces and gain access for inspection.

5.2.4.5 Bridges

Many bridges cross watercourses or other linear features providing, on their verges, commuting and foraging habitats for bats. This means that many bridges are used for roosting. Some examples are given in Billington and Norman (1997). Survey considerations that are specific to bridges are given below.

- Bats roost in many different locations within old and new bridges. Features offering potential include any holes, cracks and crevices leading to voids, particularly where there is clear access.
- Roosting locations in which bats have been recorded in bridges include expansion joints; gaps at the corner of buttresses; widening gaps; cracks and crevices between stonework and brickwork where mortar has fallen out; drainage pipes and ducts; and internal voids within box girder bridges.
- Features of bridges are given specific terms: use the technical terminology in recording and reporting. The *Bat Workers' Manual* (Mitchell-Jones and McLeish, 2004) provides useful guidance, including diagrams.
- Bridges require specific health and safety consideration because they are often associated with watercourses, roads or railway lines. Access for survey may require a boat; scaffolding; a mobile elevating work platform (MEWP); a Permit to Work; Personal Track Safety training and qualification; or a Track Visitor Permit (TVP). Survey may even require a road or rail closure. Confined spaces training may be required to access box girder bridges. All requirements should be discussed with the client and agreed with the relevant operating authority.

5.2.4.6 Underground sites

Underground sites can provide the specific microclimatic conditions that bats favour during hibernation in the winter (although they may also be used at other times of the year). A preliminary roost assessment carried out at any time of year can assess the potential for winter use, look for droppings (which can be subjected to DNA analysis for species identification) and other signs and look for bats using the site at other times of the year. However, only the winter hibernation surveys will provide information on numbers of hibernating bats. This section describes the considerations required for a preliminary roost assessment and Section 5.3 provides information on how to carry out a winter hibernation survey. The site in question may also be suitable for swarming bats; see Section 8.3 for survey methods.

- A level two class bat survey licence is required to enter known bat hibernation sites in England and in the other UK countries hibernation surveys are not included on all survey licences. It is essential that ecologists entering sites where bats are hibernating have the appropriate licence to do so.
- Ecologists entering hibernacula should be familiar with the latest information and guidance on white-nose syndrome; see Section 5.3.4 (Box 2).
- The LBG or National Bat Monitoring Programme may be aware of the site and carrying out regular monitoring already.
- It is advisable to consult mining history organisations, the BCA[41] or local caving groups before undertaking visits to natural caves and abandoned mines. These organisations frequently have important site-specific information about safety precautions, site layout, history, records of bats and details of any access agreements.
- The BCA has a Cave Conservation Code, which is downloadable from their website.[42]
- Caving groups may be available to provide training or practical assistance for survey work.
- Entering underground sites may require Confined Spaces Training or rope access. A full risk assessment should be carried out and often a method statement is also required. Equipment and training specific to the site should be identified and obtained.
- Underground sites beneath buildings, such as cellars, may be more readily accessible to ecologists than caves and mines and therefore require a different approach.

5.2.5 Complementary methods

Where bat droppings are present, samples should be carefully collected for DNA analysis (see Appendix 4 for collection protocol) unless species identification has been reliably established by other means such as observation of bats in the roost or from echolocation calls. Some species groups, for example those from the genus *Myotis* and *Plecotus*, are difficult to tell apart by these methods (Parsons and Jones, 2000; Walters *et al.*, 2012), so DNA analysis of their droppings may be necessary. DNA analysis of droppings is a more reliable method than identifying droppings by their shape, texture or colour, which can be variable and overlaps between species. Various organisations offer this service. Fabric or plastic sheets can be placed down in structures to collect droppings for this purpose on subsequent survey visits.

As a last resort, it may be possible to capture bats by hand and handle them in order to identify their species, gender and age during a preliminary roost assessment (see comments in Section 5.2.3 in relation to licensing and when handling bats is inappropriate).

5.2.6 Timing

Preliminary roost assessments can be carried out at any time of year providing any related limitations are recognised and reported.

If a maternity roost is identified, disturbance should be minimised during June and early July, when females are heavily pregnant or dependent young are present. Similarly, if a hibernation site is discovered then any subsequent disturbance should be minimised during the coldest months of December to February. Further information about these roosts can be gained from DNA analysis of bat droppings collected outside these sensitive periods (to establish species). Roost characterisation surveys (see Section 7.2) can be used to gain more information about maternity roosts and hibernation visits should be kept to a minimum (see Section 5.3).

5.2.7 Survey effort

The time needed for a preliminary roost assessment will vary according to the complexity of the structure and the number of ecologists deployed. Large structures with multiple roof spaces, multiple human access points and/or abundant voids and crevices will clearly take some time to understand and search thoroughly. Also, structures may contain several different bat roosts of different species each with their own access point and used at different times of the year. This all adds time to the survey.

As a guide, an internal inspection of a single roof area of a four-bedroom domestic property is likely to take one ecologist (with an assistant remaining outside the loft) approximately one to two hours; an internal inspection of a traditional timber-framed farm building may take one ecologist plus assistant between four hours and one day; an internal inspection of a large complex building such as a former hospital or stately home, with numerous roof voids and buildings, is likely to take one ecologist plus assistant several days. This is, of course, heavily dependent on the individual situation.

It is often difficult to have confidence in negative preliminary roost assessment survey results. For example, evidence of bats can be weathered away or bats could roost in inaccessible cracks and crevices, leaving little or no external evidence. It may therefore be necessary to spend more time searching and employ equipment such as mirrors and endoscopes.

5.2.8 Weather conditions

Preliminary roost assessments can be carried out under any weather conditions providing the survey is safe and any related limitations are recognised and reported.

5.2.9 Next steps

Where the possibility that bats are present cannot be eliminated or evidence of bats is found during a preliminary roost assessment, then further surveys (such as winter hibernation (Section 5.3), presence/absence (Section 7.1) and/or roost characterisation (Section 7.2) surveys) are likely to be necessary if impacts on the roosting habitat (or the bats using it) are predicted. The ecologist should consider the further surveys needed (if any), their logistics (resources, emergence survey locations, timings), and any potential health and safety hazards reported.

If the structure has been classified as having low suitability for bats (see Table 4.1), an ecologist should make a professional judgement on how to proceed based on all of the evidence available.

[41] http://british-caving.org.uk/
[42] http://british-caving.org.uk/wiki3/doku.php?id=conservation_access:cave_conservation_code

If sufficient areas (including voids, cracks and crevices) of a structure have been inspected and no evidence found (and is unlikely to have been removed by weather or cleaning or be hidden) then further surveys may not be appropriate. Information (photographs and detailed descriptions) should be presented in the survey report to justify this conclusion and the likelihood of bats being present at other times of the year estimated. If there is a reasonable likelihood that bat roosts could be present, and particularly if there are areas that are inaccessible for survey, then further surveys may be needed and these should be proportionate to the circumstances (see Section 2.2.5).

If no suitable habitat for bats is found, then further surveys are not necessary. In this scenario, it is necessary to document how this decision has been reached; photographs and detailed descriptions should be made available as evidence of a robust survey and assessment.

5.3 Winter hibernation surveys – structures

5.3.1 Description and aims

A **winter hibernation survey** includes a detailed inspection of a structure during the winter to look for and identify hibernating bats or other evidence of bat occupation. This survey will be necessary if potential has been identified for a structure to support hibernating bats (during the preliminary ecological appraisal (Chapter 4) or preliminary roost assessment (Section 5.2)) and the structure is likely to be impacted by proposed activities.

It should be noted that sites used for hibernation may also be used by bats at other times of the year and therefore other surveys may also be necessary.

5.3.2 Equipment

Generic documentation/equipment required for field surveys for bats is provided in Section 2.5.2; survey-specific equipment is listed in Appendix 1.

5.3.3 Expertise and licences

Section 2.5.1 discusses expertise and Section 1.2.2 provides information on licences. A winter hibernation survey could cause disturbance to bats and therefore it is good practice for ecologists to hold a survey licence. Standard survey licences for hibernacula do not permit handling of hibernating bats and this is only rarely permitted by a specific project licence. The handling of hibernating bats should therefore be avoided[43] except in the event of an emergency where the bat is in danger.

Training relating to health and safety may also be required for hibernation surveys; examples include the safe use of ladders or confined spaces training (see Section 2.7).

5.3.4 Methods

This type of survey requires close and systematic inspection of all cracks, crevices and voids for hibernating bats using torches, mirrors and endoscopes. With the exception of horseshoe bats, which usually hang freely from the walls and ceilings of hibernacula, hibernating bat species are often under-recorded because they crawl deep into crevices and can be difficult to find. Their presence is sometimes given away by droppings or oil staining around cracks and crevices or droppings beneath.

Bats periodically arouse to drink, as well as to feed if it is warm enough for insects to be active. Arousal may also be triggered by disturbance through increased levels of noise, light or heat, which may result from the presence of ecologists (therefore the number of ecologists and the amount of time they are present should be minimised). The disturbance is not always obvious to the observer at the time, as bats do not necessarily arouse immediately. There is evidence that the longer the bats have been in a torpid state, the more sensitive they are to arousal stimuli (Thomas, 1995). Bats should therefore be identified with minimal disturbance. The location and species (or genus) of all bats should be marked on a map of the structure.

Identification can be challenging because often only part of the bat can be seen. Experience is essential to gain as much information in as short a time as possible. If it is only possible to identify the bats to genus level (for example, with the *Myotis* species) then it may be possible to gain positive identification through other methods such as DNA analysis of droppings or collection of acoustic data (see Section 5.3.5).

The presence of any significant accumulations of droppings, *Nycteribiid* pupal cases (Hutson, 1984) and stained or marked areas should be recorded, as these may indicate the presence of large numbers of bats at other times of the year. Further visits during different seasons may be required in such situations to assess use of the site.

Ecologists entering hibernacula should familiarise themselves with the latest information on white-nose syndrome, provided in Box 2 (below).

[43] Handling hibernating bats has been shown to have a detrimental effect (Speakman *et al.*, 1991).

Box 2 White-nose syndrome and bats in the UK.

White-nose syndrome (WNS) is a disease caused by the fungus *Pseudogymnoascus destructans.* It affects hibernating bats in eastern North America, where it has caused the death of millions of bats since it was first discovered in 2006. Symptoms of WNS are:

- visible white fungus *(P. destructans)*, around the nose, ears, wings and/or tail membrane;
- bats clustered near the entrances of hibernacula, or in areas not normally identified as winter roost sites;
- bats flying outside during the day in temperatures at or below freezing; and
- dead or dying bats in or near hibernation sites.

Whilst the fungus associated with the syndrome has been identified on bats from at least 15 European countries since 2009, none of the other symptoms have been recorded and therefore there is no WNS in Europe (the fungus may have evolved but UK bats have an immunity that the North American species affected do not).

The fungus has been isolated from two live bats in the UK and from a number of environmental samples but as with the rest of Europe there is no evidence of WNS. BCT provides guidance for bat workers undertaking hibernation surveys and surveyors should remain vigilant and report any suspected cases of either the fungus or WNS to BCT and observe appropriate decontamination procedures. For more information refer to the WNS pages on the BCT website.[44]

5.3.5 Complementary methods

See Section 5.2.5 and Appendix 4 regarding the collection of droppings for DNA analysis. This can be particularly useful in situations where species identification is not possible because bats are tucked too far into crevices for ecologists to see their diagnostic features.

Deploying automated/static bat detectors can be useful in gaining information about hibernating bats (although the echolocation calls of *Myotis* species are notoriously difficult to separate (Parsons and Jones, 2000; Walters *et al.*, 2012)). Because the detectors can be left for long periods of time they are more likely to pick up bats when they become active, which may be particularly useful at sites with deep crevices that cannot be inspected. Detectors should ideally be deployed with temperature and humidity loggers to provide context (in terms of environmental conditions) for the survey results collected.

5.3.6 Timing

A survey at any time of year may indicate the suitability of a site for hibernation and the presence of droppings only will confirm that the site is used by bats (although an absence of droppings does not confirm absence) but further surveys may be required to determine when and how bats use the site.

The period during which bats hibernate in any given winter depends on factors such as ambient temperature, humidity and species. Some species, notably barbastelle and brown long-eared, may only hibernate for extended periods when temperatures fall below freezing. Bats can hibernate any time between November and March, depending on the prevailing weather conditions and location. Different sites are likely to be used at different times, dependent on the types of conditions they offer.

The highest numbers of bats in underground hibernacula are usually found in January. During the winter, individual bats move around to sites that present the optimum environmental conditions for their age, sex and body weight. Many species are only found in underground sites when the weather is particularly cold and therefore surveys to detect bats are most appropriate from December to February.

5.3.7 Survey effort

Because winter surveys may disturb hibernating bats, visits should be limited to the minimum necessary to gain the required information. If it is necessary to assess the numbers of bats using a site, a minimum of two visits is recommended, one in mid-January and one in mid-February.

Absence is more difficult to demonstrate and, in some cases, it may be prudent to assume that a suitable site underground in good habitat and close to other known roost sites is used by bats.

Automated/static surveys for winter activity within structures with a moderate to high likelihood of bats being present should be undertaken for a minimum of two weeks in each month from December to February.

5.3.8 Weather conditions

As the highest numbers of bats are found in the coldest conditions, it is advisable for surveys to be carried out when the weather is at its coldest.

5.3.9 Next steps

Where bat hibernation roosts are likely to be impacted by proposed activities, it will be necessary to carry out an impact assessment and design an appropriate mitigation strategy with habitat enhancements for bats where appropriate. This information is essential to inform a planning application or EPS licence application to allow the proposed activities to proceed legally.

[44] http://www.bats.org.uk/pages/about_bats-white-nose_syndrome-586.html

Chapter 6

Bat roost inspection surveys – trees

6.1 Introduction

This chapter provides information on carrying out inspection surveys for bat roosts in trees. Alternative sources of information are BS 8596:2015 Surveying for bats in trees and woodland (BSI, 2015) and the Bat Tree Habitat Key (Andrews, 2013).

These surveys may be required where development proposals include tree felling or lopping where bats or their roosts could be *directly* impacted if present.

Some of these surveys may also be needed where bats roosting in a tree could be *indirectly* impacted by development activities such as lighting or removal of vegetation.

The above principles apply regardless of the size of the development.

Surveying trees for bat roosts can be more challenging than surveying buildings because many species that use trees for roosts are known to exhibit roost switching behaviour, including barbastelle, Bechstein's bat, Daubenton's bat, Natterer's bat, Leisler's bat, noctule, common pipistrelle and brown long-eared bat (Harris and Yalden, 2008, Dietz *et al.*, 2011). Some UK examples are as follows: Smith and Racey (2008) observed roost switching in Natterer's bat on average every 3 days; and Waters *et al.* (1999) observed roost switching in Leisler's bat between every 2 and 10 days. Frequent roost switching has also been observed in barbastelle (Billington, 2003; Greenaway, 2001; Zeale, 2011) and Bechstein's (Palmer, 2013), two of our rarest species.

Additional difficulties inherent in finding tree-roosting bats are as follows: droppings do not persist in trees in the same way as they do in buildings; some tree-roosting bats echolocate very quietly (and sometimes not at all) and are therefore difficult to detect using bat detectors; some tree-roosting bats emerge from their roosts very late and return very early; and emergence surveys are often constrained due to the height of tree roosts above ground level and restricted observation due to foliage or lack of light under the canopy. The chances of discovering a roost, even if one is present, are relatively low. However, some of our rarest species are heavily reliant on tree roosts.

Due to these limitations and from what is known about the ecology of tree-roosting bats, it is arguable that all trees with bat roosting potential should be considered part of a resource that will be used at one time or another by tree-roosting bats in order to determine the extent of impacts. Survey work on individual trees may confirm presence but is unlikely to conclusively confirm absence. Precautionary measures are likely to still be essential during works even where surveys have not identified occupancy.

Where survey work is required, it should be designed to answer specific questions, such as:

- Are actual or potential bat roosts present (and where are they)?
- Which bat species use the site for roosting?
- How many bats are these roosts likely to support?
- What is the current arrangement of vegetation and lighting in relation to the access points?
- At what times of the year are bats present? How does use change seasonally?
- What types of bat roost are present, e.g. day, night, feeding, transitional/occasional, maternity, hibernation, satellite (see Section 3.3).

Answering some or all of these questions allows an ecologist to carry out an impact assessment and design a mitigation, enhancement and monitoring strategy, where relevant.

Roost surveys of trees generally take a staged approach, with the first step being a **preliminary ground level roost assessment** (possibly preceded or combined with a preliminary ecological appraisal; see Chapter 4), which may be followed up by **PRF inspection, presence/absence** and/or **roost characterisation surveys**. The latter two survey types are covered in Chapter 7, which also covers structures. Survey design should be iterative; each stage informing the next, as per the flow chart provided in Figure 6.1. The effectiveness of the surveys should be considered at each stage.

Figure 6.1 Flow chart illustrating the process used to establish which types of survey are necessary for roosts in trees.

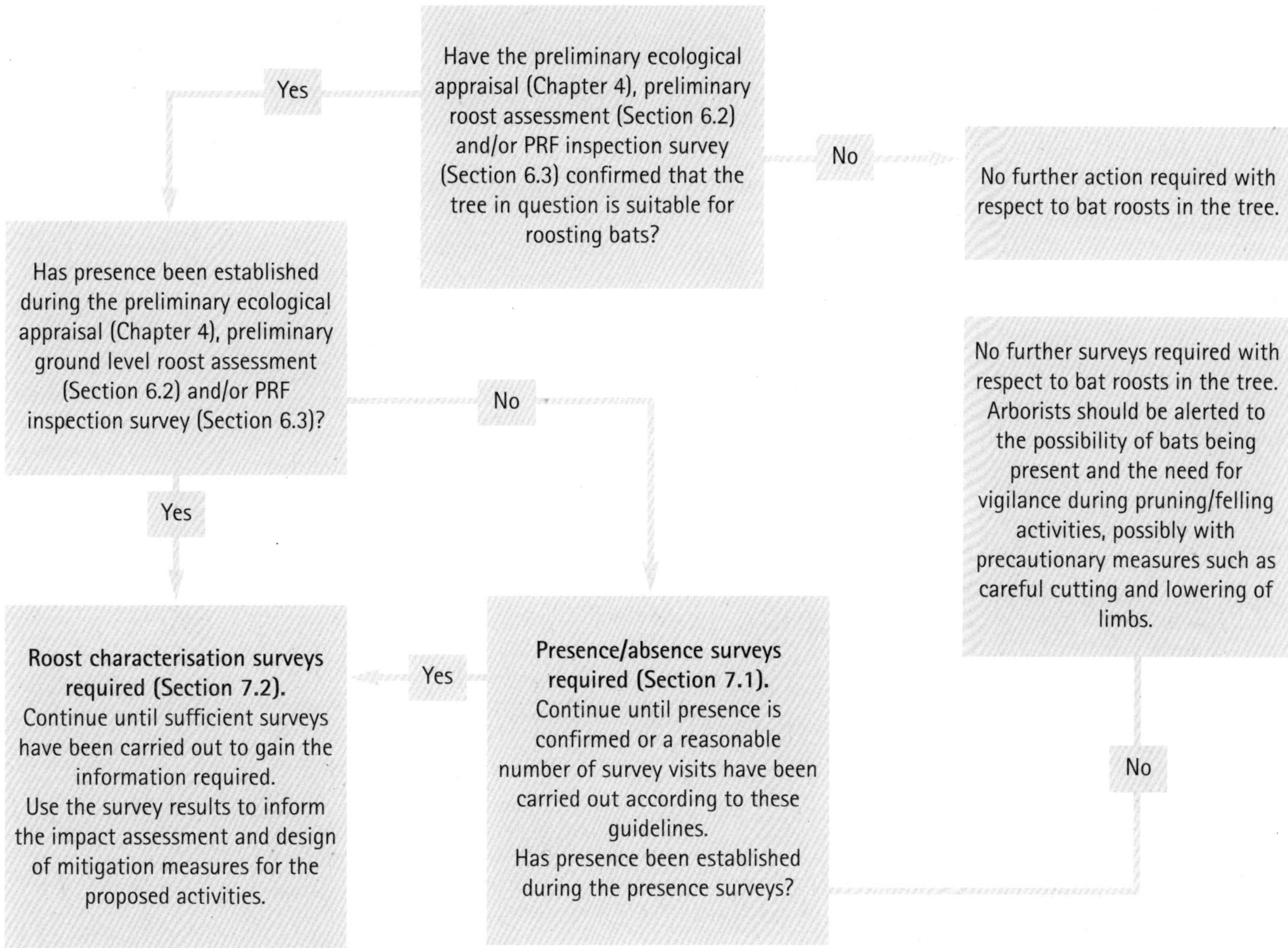

6.2 Preliminary ground level roost assessment – trees

6.2.1 Description and aims

A **preliminary ground level roost assessment** of a tree is a detailed inspection of the exterior of the tree from ground level to look for features that bats could use for roosting (PRFs). The aim of this survey is to determine the actual or potential presence of bats and the need for further survey and/or mitigation.

6.2.2 Equipment

Generic documentation/equipment required for field surveys for bats is provided in Section 2.5.2; survey-specific equipment is listed in Appendix 1.

6.2.3 Expertise and licences

Section 2.5.1 discusses expertise and Section 1.2.2 provides information on licences. A preliminary ground level roost assessment of trees is unlikely to result in disturbance to bats unless the ecologist intends to investigate low-level PRFs in trees with a torch or endoscope. If disturbance to bats is a possibility, then a survey licence is required.

6.2.4 Methods

The method involves a detailed inspection of the tree from ground level to compile information about the tree, PRFs (or lack of), and evidence of bats. Sufficient time should be allowed to complete the inspection during daylight hours. Poor light conditions can mean that PRFs are missed in trees. The inspection should be carried out systematically and consistently around all parts of the tree (from all angles and from both close to the trunk and further away) and the results recorded in a standard format. High-level PRFs can be identified by shining bright torches on cavities and shaded areas of the branches and using binoculars can help to focus in more detail.

All trees surveyed should be numbered and marked on a map or plan of the site (in some situations even trees with no PRFs should be mapped as a record). Information collected about the tree should at least include the location (grid reference) and tree species. Diameter at breast height can also be measured using a specialist tree tape (logger's tape) or number of stems can be recorded if the tree has been coppiced. This information will enable ecologists to locate the tree on subsequent visits. It is often difficult to find trees in a group or in woodland on a second survey visit and therefore marking individual trees with a tag or some tape may be essential. The permission of the landowner should be sought for this.

PRFs that may be used by bats include:

- woodpecker holes;
- rot holes;
- hazard beams;

- other vertical or horizontal cracks and splits (such as frost-cracks) in stems or branches;
- partially detached platey bark;
- knot holes arising from naturally shed branches, or branches previously pruned back to the branch collar;
- man-made holes (e.g. cavities that have developed from flush cuts) or cavities created by branches tearing out from parent stems;
- cankers (caused by localised bark death) in which cavities have developed;
- other hollows or cavities, including butt-rots;
- double-leaders forming compression forks with included bark and potential cavities;
- gaps between overlapping stems or branches;
- partially detached ivy with stem diameters in excess of 50mm;
- bat, bird or dormouse boxes.

Andrews (2013) provides more information on specific arboricultural terms for these features and how/why they form in trees.

Information collected about PRFs should include a description, the height of the feature above ground level, the orientation of the feature in relation to the trunk and the orientation of the access to the feature. This information will enable ecologists to locate the PRF on subsequent visits.

Signs of a bat roost, besides the actual presence of bats, include:

- bat droppings in, around or below a PRF;
- odour emanating from a PRF;
- audible squeaking at dusk or in warm weather;
- staining below the PRF.

Some of these signs (odour, squeaking) may be the result of other animals such as birds or squirrels and staining may be the result of wet rot, which would preclude roost presence. Bats or bat droppings are the only truly conclusive evidence of a roost but many bat roosts have no external signs.

During a preliminary ground level roost assessment of trees a more precise assessment of suitability is made than during a preliminary ecological appraisal (see Table 4.1 on page 35 for definitions of suitability). However, the evaluation at this stage is still relatively basic because it is not possible to inspect PRFs (except those at ground level) more closely to ascertain their true potential for supporting roosting bats. A tree should be categorised according to the highest suitability PRF present.

6.2.5 Complementary methods

See Section 5.2.5 and Appendix 4 regarding the collection of droppings to enable identification using DNA analysis. The main constraint with respect to collecting droppings from trees is their quality, because droppings can rapidly decay in trees.

6.2.6 Timing

Preliminary ground level roost assessments of trees are best carried out in winter (after the leaves have fallen and before new ones replace them – around December to March). If it is necessary to carry out these surveys when the leaves are on the trees, then it may not be possible to see all PRFs and surveys may need to be repeated in the winter months or a more thorough PRF inspection survey carried out to detect all PRFs, as far as possible. When these surveys are carried out in the summer, it may be possible to hear bats making audible social calls (or non-audible calls, using a bat detector) from roosts in trees. An example is available on the CD-ROM that accompanies *Woodland Management for Bats* (FC England *et al.*, 2005).

6.2.7 Survey effort

The time needed for a preliminary ground level roost assessment will vary according to the size of the trees, the number of PRFs and the number of ecologists deployed.

As a guide, it may be possible for a single ecologist to inspect 20–30 trees in a day if those trees are large, veteran oaks with multiple PRFs. It may, however, be possible to inspect double the number or more if the trees are smaller and with less potential for roosting bats.

6.2.8 Weather conditions

Preliminary ground level roost assessments for trees are best carried out in bright, dry and calm weather because these conditions maximise the chances of seeing PRFs.

6.2.9 Next steps

Where suitable roosting habitat (moderate or high suitability; see Table 4.1 on page 35) or evidence of bats is found during a preliminary ground level roost assessment then further surveys (such as PRF inspection surveys (Section 6.3), presence/absence surveys (Section 7.1) or roost characterisation surveys (Section 7.2)) are likely to be necessary if impacts on the roosting habitat or the bats using it are predicted. The ecologist should consider the further surveys needed (if any), their logistics (resources, emergence survey locations, timings), and any potential health and safety hazards reported.

If no or low suitability PRFs for bats are found (using the definitions in Table 4.1) then further surveys are not necessary. In this scenario, it is necessary to document how this decision has been reached; photographs and detailed descriptions should be made available as evidence of a robust survey and assessment. Where there are low suitability PRFs, precautionary measures may be appropriate during felling or pruning activities.

If ground level surveys are inconclusive, and PRFs could be present at height, it may still be necessary to carry out further surveys (see Section 6.3).

6.3 PRF inspection surveys – trees

6.3.1 Description and aims

A **PRF inspection survey** involves the use of tree-climbing or access equipment such as cherry pickers, MEWPs or scaffold towers to gain access to PRFs to assess in more detail their likely suitability for bats and to look for evidence of bats such as live or dead bats, droppings, staining or odour. These surveys are valuable to prevent unnecessary emergence/dawn work where features appear to be of high suitability from the ground but are actually of limited or no suitability. Tree climbing is often the most effective way to access all features but may be constrained by health and safety issues (e.g. trees may be unsafe to climb) and therefore it may be more appropriate to use alternative access equipment or skip to presence/absence surveys (see Section 7.1).

The aim of this survey is to reclassify PRFs and determine the presence/absence of bats at the time of the survey and the need for further survey and/or mitigation.

6.3.2 Equipment

Generic documentation/equipment required for field surveys for bats is provided in Section 2.5.2; survey-specific equipment is listed in Appendix 1.

6.3.3 Expertise and licences

Section 2.5.1 discusses expertise and Section 1.2.2 provides information on licences. A PRF inspection survey to look for bats could cause disturbance and therefore it is good practice for ecologists to hold a survey licence. Where bats are present, this allows immediate identification, reducing the risk that the bats will remain unidentified if not present on a subsequent visit.

In order to carry out PRF inspection surveys using tree climbing, ecologists should be trained, qualified and experienced in tree climbing and aerial rescue and only work in pairs. Skills should be kept up-to-date through regular use and refresher courses should be considered for those who use these skills only infrequently. In this scenario, it may be appropriate for an ecologist to team up with an arborist to ensure that surveys are carried out as safely and efficiently as possible.

Owner- or operator-specific training may also be required when ecologists employ cherry pickers, MEWPs or scaffold platforms to access PRFs for inspection.

6.3.4 Methods

The method involves accessing PRFs using a harness and ropes (or other access equipment) to carry out a detailed internal inspection using torches, mirrors and endoscopes to compile information on the dimensions and protection from the elements and to search for evidence of bats. PRFs that appear to be of high suitability from the ground may only be of low suitability because, for example, they are filled with rainwater. Close inspection of features can be extremely useful because it facilitates a much more reliable assessment of suitability and provides an opportunity for bats and bat droppings to be found if they are present.

Sufficient time should be allowed to complete PRF inspection surveys during daylight hours. Poor light conditions could jeopardise safety and cause disturbance to bats at emergence time. The inspection should be carried out systematically and consistently around all parts of the tree and the results recorded in a standard format.

During a PRF inspection survey, the ecologist should collect information about the dimensions of features as this information may be required at a later stage. The ecologist should also review the evaluation that was made during the preliminary ground level roost assessment (see Section 6.2) according to the definitions provided in Table 4.1 on page 35. The evaluation at this stage is more accurate due to PRFs being more closely inspected. A tree should be classified according to the highest suitability PRF identified during the tree climbing survey.

6.3.5 Complementary methods

See Section 5.2.5 and Appendix 4 regarding the collection of droppings to enable identification using DNA analysis. The main constraint with respect to collecting droppings from trees is their quality, because droppings can rapidly decay in trees.

6.3.6 Alternative methods

Where there are large numbers of trees, the efficiency and efficacy of PRF inspection and other techniques should be evaluated and alternative methods considered. In situations where there are a lot of trees to survey, such as in woodland, it may be more effective to consider advanced licence bat survey techniques (ALBST) such as trapping and radio tracking to locate tree roosts. Such methods are invasive and can be expensive, therefore the decision to use them should be led by the potential impacts of the proposals and thus the requirement to collect the data. ALBST are covered in Chapter 9.

6.3.7 Timing

PRF inspection surveys can be carried out at any time of year, although the likelihood of discovering evidence of bats at different times should be considered.

Tree climbing surveys should also consider other protected species such as birds and red squirrels and, if present, the timing of surveys may need to be adjusted accordingly or a specific licence may be required.

6.3.8 Survey effort

The time needed for PRF inspection surveys will vary according to the size of the trees and the number of PRFs. For tree climbing, time taken often depends on experience. Efficiency can be gained by teaming up ecologists with arborists, who are often more experienced in accessing difficult areas of trees. For PRF inspection surveys using access equipment such as cherry pickers, the time required is likely to depend more on ground conditions and barriers to movement such as hedgerows.

As a guide, it may be possible for an ecologist to inspect only two to four trees in one day if those trees are large, veteran oaks with multiple PRFs. It may, however, be possible to inspect two or three times this many if the trees are smaller and with less potential for roosting bats.

Andrews and Gardener (2015) presented a summary of evidence and an encounter probability model for PRF inspections for tree-roosting bats. The model suggests that a very high number of visits is required to be sure of encountering bats; likely survey 'success' needs to be taken into account when designing surveys to capture evidence of bats and interpreting their findings.

6.3.9 Weather conditions

Tree climbing surveys are best carried out in dry and calm weather for safety reasons.

6.3.10 Next steps

Where a PRF has been verified as moderate or high suitability for bats or evidence of bats is found, further surveys are likely to be necessary if impacts on the PRF or

the bats using it are predicted (Section 7.1 and 7.2). These are particularly important where features could not be inspected at all; could not be inspected in their entirety because they were too extensive; or where evidence of bats may have been removed by the weather or invertebrates resident in the PRF. The ecologist should consider the further surveys needed (if any), their logistics (resources, emergence survey locations, timings), and any potential health and safety hazards reported.

If no or only low suitability PRFs for bats are found then further surveys are not necessary. In this scenario, it is necessary to document how this decision has been reached: photographs and detailed descriptions should be provided to the client as evidence that an adequate survey has been carried out and the conclusions are reasonable. Where there are low suitability PRFs precautionary measures may be appropriate during felling or pruning activities.

Chapter 7

Emergence/re-entry surveys – structures and trees

7.1 Presence/absence surveys

7.1.1 Description and aims

Presence/absence surveys include dusk and/or dawn visits to watch, listen for and record bats exiting or entering bat roosts. If the presence of bats has been confirmed, then roost characterisation surveys (see Section 7.2) may be required (depending on how much information on species, numbers, access points, roosting locations, timing of use and type of roost has already been collected), although other features, structures or trees on site may still require presence/absence surveys.

Presence/absence surveys would be needed if:

- the preliminary roost assessment (structures and trees) has not ruled out the reasonable likelihood of a roost being present (because there are locations with potential for bats to roost undetected in concealed cracks, crevices or voids), but no definitive evidence of the presence of bat roosts has been recorded;
- the PRF inspection survey (trees) has identified moderate and high suitability PRFs for bats but no definitive evidence of the presence of bat roosts has been recorded;
- a comprehensive inspection survey is not possible because of restricted access, but there are features with a reasonable likelihood of supporting bats; and/or
- there is a risk that evidence of bat use may have been removed by weather or human activities.

The aim of this survey is to determine the presence or absence of bats at the time of the survey and the need for further survey and/or mitigation.

The additional limitations of tree surveys (in comparison to surveys of structures) are highlighted in Section 6.1.

7.1.2 Equipment

Generic documentation/equipment required for field surveys for bats is provided in Section 2.5.2; survey-specific equipment is listed in Appendix 1.

7.1.3 Expertise and licences

Section 2.5.1 discusses expertise and Section 1.2.2 provides information on licences. Presence/absence surveys are unlikely to disturb bats if carried out correctly; however, it is good practice for these surveys to be designed and carried out, or at least led, by licensed surveyors who have gone through a period of training and evaluation.

7.1.4 Methods

The method involves ecologists visiting at dusk or dawn to listen/record (using a bat detector) and watch for bats emerging or returning to roosts and compile information on species, numbers, access points and roosting locations. This should be informed by the preliminary roost assessment (see Sections 5.2 and 6.2), which identified potential roosting and access points, and by the PRF inspection survey for trees (see Section 6.3), which clarified the potential suitability of different PRFs to bats. These places should be the focus of the survey and their number and arrangement should inform the number and arrangement of surveyors required to complete the survey (although ecologists should be aware that bats may emerge in unexpected places). Ecologists should be adequately briefed about the exact area they are expected to observe for emerging or returning bats and the areas their colleagues are observing to avoid double-counting. Radio contact can help ecologists to communicate easily and quickly about their observations.

Sufficient surveyor coverage of a structure is required and it is important that enough ecologists are used to thoroughly observe all potential access points, ideally during a single survey, and this should be checked by those assessing surveys and reports.

Generally, one ecologist can only observe two sides of a simple structure, from the corner, and their ability to do so reduces as the complexity and size (i.e. length/width) of the structure increases or where observation is obscured by a tall hedge, wall or other obstacle. More complex structures or multiple structures require more ecologists, particularly if there are many potential access points, as all areas with potential should be covered. If fewer ecologists are available it may be necessary to visit the site (standing at different locations each time) over several consecutive nights (collectively considered to be one survey visit) to cover all areas.

Ecologists should consider whether it will be possible to watch all the PRFs on a tree with a single ecologist and use additional ecologists where necessary, for example where PRFs are on different aspects of a tree or one or more PRFs are obscured by foliage. It is sometimes not possible to see all PRFs from the ground so this should also be taken into account.

It may be possible to use fewer ecologists to watch for bats exiting; for example, a block of buildings or a woodland as a whole unit, but this would only identify that roosts were present within the block/woodland and would not identify individual buildings, trees or roosts. The choice of method depends on the amount of detail required to meet the survey aims.

Surveyors should be stationary to avoid bats being missed. One or two ecologists walking around a large site are unlikely to pick up individual bats or small roosts and could even miss larger roosts and is not appropriate.

Ecologists should concentrate and maintain visual contact with the relevant access points throughout (this can be facilitated by using a voice recorder) because single or small numbers of bats can emerge very quickly and are difficult to observe, particularly as light levels decrease at dusk (and they do not always echolocate). Where possible, ecologists should stand close enough to the relevant access points to be able to identify late-emerging, quiet-calling bat species (see Section 3.9). Some species are only detectable to a few metres and emerge in darkness (a torch should not be used). Dawn surveys may be more effective in this situation (and where there are only small roosts) because when bats return to the roost at dawn they often fly around outside, and may repeatedly land on roost access points prior to entering, whereas at dusk they often emerge and immediately fly away. Dawn surveys can be particularly useful for trees.

In some situations, for example with large open barn doors, it may be more effective for the ecologist to stand inside the doorway looking out to observe emerging bats against the lighter night sky rather than to stand outside the doorway looking into darkness.

If bats are observed emerging from structures, this does not necessarily mean they are roosting in the same location as the exit point; it may be necessary to identify roosting locations separately. Sometimes this can be established during the preliminary roost assessment. Survey design should be iterative, each survey informed by the previous one.

The results of the surveys should be recorded in a standard format using a pre-designed survey form.

7.1.5 Complementary methods

Night-vision scopes or infrared or thermal imaging cameras can increase precision in presence/absence surveys because bats are less likely to be missed if the camera is pointed at the relevant access point. This can be particularly important where there is potential for late-emerging species (see Section 3.9) and in dark conditions (for example, under the tree canopy and among fluttering foliage). Where footage is recorded, this can be analysed afterwards. However, the limited field of view offered by many systems should be considered if multiple exit points need to be observed. Infrared systems also require a separate source of true infrared illumination (not a red light filter) to be effective. While such equipment is very useful as a complementary technique, it should not be used to replace surveyors to any significant degree; the majority of any site should be observed by surveyors.

Deploying automated/static bat detectors inside a structure can be particularly useful in gaining information about late-emerging species that often fly around inside the roost prior to emergence. Caution should be exercised in using automated/static detectors for this purpose, however, because sometimes they can detect bats flying outside a structure, not just those flying inside.

7.1.6 Alternative methods

See Section 6.3.6 for alternative methods to detect the presence of bats in trees.

7.1.7 Timing

Recorded bat activity is dependent on the prevailing conditions at the time of the survey, which vary temporally (through the night, between nights, through the seasons and between years) and spatially (dependent on latitude and longitude).

Bat activity is also determined by what the bats are doing at different times of the year (although this is also dependent in part on prevailing conditions); the bat life cycle is given in Section 3.2.

The bat active period is generally considered to be between April and October inclusive (although the season is likely to be shorter in more northerly latitudes). However, because bats wake up during mild conditions in the winter to drink, feed and change roost, bat activity can also be recorded during the winter months (winter hibernation surveys of structures are covered in Section 5.3).

In general:

- April surveys may detect transitional roosts.
- May to August surveys may detect maternity colonies and males/non-breeding females in summer roosts.
- August is particularly good for maximum counts of both adults and juveniles and can be useful to observe roost re-entry because the young bats are inexperienced at flying and are often easy to observe as they try to enter the roost.
- August to October surveys may detect mating bats.
- September and October surveys may detect transitional roosts used after bats have dispersed from maternity colonies but before they go into hibernacula (although October may be less suitable for surveys in more northerly latitudes).

It is important to stress that prevailing conditions and local trends in bat activity (for example, when were the young born in the year in question?) should be considered and recorded to provide context to survey results.

Surveys should be designed around the information that is required to achieve the survey aims. Recommended timings for surveys are given in Table 7.1 below. This should be adjusted (earlier or later) if necessary by the ecologist, bearing in mind the site-specific circumstances, although this should be fully justified in the survey report.

Please note that these are the timings recommended for presence/absence surveys. Some roost characterisation surveys (see Section 7.2.7) may be appropriate in April (to identify transitional roosts) and October (to identify transitional and mating roosts) depending on the findings of previous surveys, the weather and the location (although please note that October surveys are not considered appropriate in Scotland).

Table 7.1 Recommended timings for presence/absence surveys to give confidence in a negative result for structures (also recommended for trees but unlikely to give confidence in a negative result).

Low roost suitability	Moderate roost suitability	High roost suitability
May to August (structures) No further surveys required (trees)	May to September[a] with at least one of surveys between May and August[b]	May to September[a] with at least two of surveys between May and August[b]

[a] September surveys are both weather- and location-dependent. Conditions may become more unsuitable in these months, particularly in more northerly latitudes, which may reduce the length of the survey season.
[b] Multiple survey visits should be spread out to sample as much of the recommended survey period as possible; it is recommended that surveys are spaced at least two weeks apart, preferably more, unless there are specific ecological reasons for the surveys to be closer together (for example, a more accurate count of a maternity colony is required but it is likely that the colony will soon disperse). If there is potential for a maternity colony then consideration should be given to detectability. A survey on 31 August followed by a mid-September survey is unlikely to pick up a maternity colony. An ecologist should use their professional judgement to design the most appropriate survey regime.

Different species vary in the time they tend to emerge and return to the roost according to their flight and predator avoidance capabilities. *Pipistrellus* species and noctule often emerge early and return late; brown long-eared bat and Natterer's bat often emerge late and return early (see Section 3.5).

Table 7.2 gives recommended timings for dusk and dawn surveys. These are times that ecologists should be in place.

Table 7.2 Recommended timings for presence/absence surveys.

Survey type	Start time	End time
Dusk emergence	15 minutes before sunset[a]	1.5–2 hours after sunset[b]
Dawn re-entry	1.5–2 hours before sunrise[b]	15 minutes after sunrise[c]

[a] Survey start time should be adjusted on subsequent surveys if bats are recorded already in flight at 15 minutes before sunset on the first survey (or, if only one survey had been planned, this survey may then need to be repeated).
[b] The possibility of late-emerging and early-returning species should be considered in setting times for surveys (see Section 3.5).
[c] If bats are still in flight 15 minutes after sunrise then ecologists should remain in position until all the bats have entered their roosts.

Although these time periods mean that some of the survey is in complete darkness, ecologists can still listen out for and record activity and may be alerted to the possible presence of a roost of late-emerging species so that survey methods can be adjusted either at the time or on a subsequent survey. Adjustments could include changing to a dawn survey; using night-vision scopes or infrared or thermal imaging cameras at dusk or dawn; or deploying an automated/static detector inside a structure.

Other considerations in terms of timing are as follows:
- if a roost emergence point is not lit by the setting sun, it is likely to be darker and bats may emerge earlier and return later;
- if bats have vegetation cover close to the roost they may emerge earlier and return later because the vegetation offers protection;
- if there have been periods of prolonged bad weather bats may adjust their behaviour to increase foraging times by emerging earlier or returning later;
- poor weather conditions may cause bats to alter their emergence/return times (see Section 2.6.1); and
- if the roost is very large some of the bats may emerge earlier and return later.

Timings may be adjusted (earlier or later) if necessary by the ecologist, bearing in mind the site-specific circumstances, although this should be fully justified in the survey report.

7.1.8 Survey effort

More ecologists with more equipment (if used correctly) in more seasons and under the right weather conditions generally increases the likelihood of discovering bats. However, surveys should always be proportionate to the circumstances, which can only be assessed using professional judgement.

Table 7.3 provides the minimum recommended numbers of survey visits to give confidence in a negative result for structures. Confidence in a negative result is not possible for trees due to limitations outlined in Section 6.1. The number of visits could be adjusted (up or down) if necessary by the ecologist, bearing in mind the site-specific circumstances, although this should be fully justified in the survey report.

Table 7.3 Recommended minimum number of survey visits for presence/absence surveys to give confidence in a negative result for structures (also recommended for trees but unlikely to give confidence in a negative result).

Low roost suitability	Moderate roost suitability	High roost suitability
One survey visit. One dusk emergence or dawn re-entry survey[a] (structures). No further surveys required (trees).	Two separate survey visits. One dusk emergence and a separate dawn re-entry survey.[b]	Three separate survey visits. At least one dusk emergence and a separate dawn re-entry survey. The third visit could be either dusk or dawn.[b]

[a] Structures that have been categorised as low potential can be problematic and the number of surveys required should be judged on a case-by-case basis (see Section 5.2.9). If there is a possibility that quiet calling, late-emerging species are present then a dawn survey may be more appropriate, providing weather conditions are suitable. In some cases, more than one survey may be needed, particularly where there are several buildings in this category.

[b] Multiple survey visits should be spread out to sample as much of the recommended survey period (see Table 7.1) as possible; it is recommended that surveys are spaced at least two weeks apart, preferably more. A dawn survey immediately after a dusk one is considered only one visit.

Some situations may justify a dawn survey being carried out the morning after a dusk survey. For example, if it is not clear exactly where a bat emerged from or even that the bat actually emerged, a dawn survey can be used to clarify the situation. An ecologist will be able to adjust his/her position for the dawn survey to get a better view. This may be important if the roost is thought to be transitional because the bat may have moved on by the next survey visit. If the dusk survey is conclusive, then there is less value in carrying out a dawn survey immediately after. A dusk survey immediately followed by a dawn survey should be considered to be only one survey visit because this is insufficient time for roosting behaviour to have significantly changed.

Numbers of surveys may need to be increased from those recommended in Table 7.3 where thorough internal inspections have not been possible; the number should be decided using professional judgement and rationale reported. Internal inspections (of structures and PRFs) can provide historical evidence of bat presence whereas emergence and dawn surveys only provide information about bat presence or absence at the time of the survey.

7.1.9 Weather conditions

Please refer to Section 2.6.1 for guidance on weather.

7.1.10 Next steps

If presence of a bat roost(s) is established, the next stage of the process is to carry out roost characterisation surveys (see Section 7.2 – depending on how much information on species, numbers, access points, roosting locations, timing of use and type of roost has already been collected), although it may be necessary to continue with presence/absence surveys of other parts of the structure, tree or site.

In structures, where likely absence has been adequately established, then no further action is required in relation to bats. However, it may be appropriate for contractors to be briefed about the risk of discovering bats unexpectedly during works and the need to stop work in this scenario.

In trees, it is very difficult to have confidence that roosts are absent (see Section 6.1) and therefore, even where no bats are found, it may still be necessary to apply precautionary measures when carrying out tree felling and pruning activities.

7.2 Roost characterisation surveys

7.2.1 Description and aims

When presence is established, this should trigger **roost characterisation surveys** unless sufficient information has already been collected to inform the impact assessment and design of mitigation measures. Roost characterisation surveys include emergence/re-entry surveys. They also include the collection of information about the physical characteristics of the roost and surrounding area.

The aim of these surveys is to answer the questions outlined in Sections 5.1 and 6.1, and to ascertain the features and characteristics of the roost (for example size, perching points, aspect, orientation, temperature, humidity, lighting) and the surrounding area (for example proximity of vegetation to exit points, availability of foraging areas locally) that are important.

All of this information can then be used to assess the potential impacts of the proposed development activity and design suitable mitigation and monitoring strategies. For example, information on roost characteristics may be required to inform the construction of a like-for-like replacement roost where the original roost will be lost. This information is essential when applying for planning permission or an EPS licence.

The additional limitations of tree surveys (in comparison to surveys of structures) are highlighted in Section 6.1.

7.2.2 Equipment

Generic documentation/equipment required for field surveys for bats is provided in Section 2.5.2; survey-specific equipment is listed in Appendix 1.

7.2.3 Expertise and licences

The expertise and licences required are the same for both presence/absence surveys and roost characterisation surveys (see Section 7.1.3).

7.2.4 Methods

The method used is the same for both presence/absence surveys and roost characterisation surveys (see Section 7.1.4).

Some bat species will not waste energy echolocating in higher light levels, which means other methods should be used to gain the species identification information required; for example, DNA analysis of droppings (see Section 5.2.5 and Appendix 4) or handling of bats (see Section 7.2.5). Visual cues such as behaviour, size, wing shape and ear shape may also contribute to identification but in most cases these cannot be used in isolation.

The collation of information about the physical characteristics of the roost and surrounding area is discussed below.

❍ **Size and nature of roost**

In structures, the size of the roost, including the presence and location of timber joints and other features supporting roosts, should be documented if it is likely that a replacement roost will be required. The size and nature of the internal space may be important to bats that fly around inside prior to emerging, most notably *Plecotus*, *Rhinolophus* and some *Myotis* species. The number and location of all access points (and their dimensions, which can be important for some species) should also be documented.

In trees, the dimensions of the roost feature should have been documented during the PRF inspection survey (see Section 6.3) if it has been possible to carry one out.

❍ **Roosting surfaces**

In structures, the availability of appropriate roosting surfaces (e.g. natural materials such as wood) is a key measure of the ecological functionality of a site, and should be recorded if it is likely that the roost will need to be replaced.

❍ **Aspect and orientation**

The aspect, orientation and shading of the roost and associated access points should be carefully documented, again so that this can be replicated in a replacement roost if necessary. Aspect and orientation affect how the roost is heated by the sun, although in structures heating may also result from man-made features such as boilers. If this is the case, it should also be recorded.

❍ **Temperature and humidity**

Williams (2010) and Gunnell *et al.* (2013) state that one of the factors making structures suitable for roosts is their ability to provide a stable microclimate and that temperature plays a key role in roosting ecology and selection. Where proposals will result in the loss of a maternity or hibernation roost, the temperature and humidity inside and outside the roost should be monitored using data loggers to understand how conditions fluctuate in relation to ambient temperatures throughout the season the roost is used (although this may be constrained by limited access to the areas bats are actually using). In structures that are used by bats at different times of the year, it may be necessary to collect data during more than one season. It can be the damping of temperature variation, rather than absolute temperatures, that make a roost suitable for bats. Collecting data inside and outside the roost will help to understand this and replicate conditions, where possible, in replacement roosts. Different conditions are likely to suit different species (see, for example, Boonman, 2000; Smith and Racey, 2005; Davidson-Watts and Jones, 2006).

❍ **Lighting**

Current lighting levels and locations should be noted to provide a comparison with new lighting proposals. Even one change such as an outside security light can have an impact and lighting needs to be considered in relation to current and proposed new bat access points. In cases where no significant change is proposed, it may not be necessary to measure the light levels at all, but current lighting fixtures should be plotted.

❍ **Habitat**

Vegetation in close proximity to a roost can be extremely important for some species of bat that seek cover from predators and the weather immediately after emerging. It also provides structure for acoustic orientation and navigation and opportunities for foraging. Features likely to be important to bats should be noted so that these can be retained or replicated post-development as necessary. The importance of different habitat features vary from species to species (see, for example, Davidson-Watts *et al.*, 2006; Entwistle *et al.*, 1997).

7.2.5 Complementary methods

The complementary methods are the same for both presence/absence surveys and roost characterisation surveys (see Section 7.1.5).

It may also be possible to capture bats using a hand net in order to identify their species, gender and age during a roost characterisation survey. The correct licence (see Section 1.2.2), knowledge and skills (see Section 2.5.1) should be in place to carry out this activity and sensitive times of year should be avoided (such as when bats are heavily pregnant or with dependent young).

7.2.6 Alternative methods

See Section 6.3.6 for alternative methods to detect the presence of bats in trees.

7.2.7 Timing

See Section 7.1.7; comments on timing are the same for both presence/absence surveys and roost characterisation surveys. It may be appropriate to carry out surveys in April and/or October depending on the need to characterise transitional roosts or mating roosts, the findings of previous surveys, the weather and the location (although please note that October surveys are not considered appropriate in Scotland).

7.2.8 Survey effort

Survey effort required to collect the relevant information that is needed for an impact assessment and the design of mitigation strategies is very much site-specific. Dusk and dawn surveys should be repeated until the information outlined in Sections 5.1 and 6.1 is reliably collected, although appropriate methods and equipment should be used to minimise the number of repeat visits required and effort should always be proportionate to impact.

If presence has been confirmed by droppings found during a preliminary roost assessment (Sections 5.2 and 6.2) but bats have not been detected during roost characterisation surveys, it may be necessary to carry out further surveys at alternative times of year.

7.2.9 Weather conditions

Please refer to Section 2.6 for guidance on weather.

7.2.10 Next steps

Where bat roosts are likely to be impacted by proposed activities it will be necessary to carry out an impact assessment and design an appropriate mitigation and monitoring strategy with habitat enhancements for bats where appropriate. This information is essential to inform a planning application or EPS licence application to allow the proposed activities to proceed legally.

Chapter 8

Bat activity and back-tracking surveys

8.1 Introduction

This chapter provides information on carrying out bat detector surveys for bats. These bats may be commuting, foraging or exhibiting social behaviour (such as calling for mates during the mating season or swarming in the autumn). Acoustic surveys enable identification of species and provide an index of bat activity. Actual numbers of individuals can often not be established unless acoustic data is coupled with direct observations in the field by an ecologist, or through recordings made by an infrared or thermal-imaging camera.

These surveys may be required where development proposals are likely to impact on habitats suitable for bat commuting and foraging (see Section 2.2.2). Road and rail schemes can cause the specific impact of collision and it is good practice to carry out automated/static bat activity surveys of crossing points.

As with all surveys, survey design should be based around the questions that require answers. For the purposes of development and planning, the main questions with respect to bats in flight away from their roosts are generally as follows:

- Are bats present or absent?
- Which bat species use the site?
- What are the activity levels of bats on the site and can this tell us anything about the abundance (number) of bats using the site?
- What are bats using the site for?
- What is the temporal (both seasonally and in relation to time of night) and spatial distribution of recorded bat activity on site?
- Are peaks in bat activity associated with particular temporal and/or spatial locations, e.g. times of night or parts of the site?
- How are the habitats used on site connected to habitats in the surrounding area?

Answering some or all of these questions should allow an ecologist to carry out a robust impact assessment.

In order to answer these questions, bat activity surveys generally begin with the preliminary ecological appraisal, which includes a desk study and fieldwork (see Chapter 4). This provides existing data about bats in the area and identifies and assesses the suitability of habitats on site for bats. This information should be used to inform survey design, which should be iterative; each stage should inform the next.

The following sections describe **transect** and **automated/static bat activity surveys, back-tracking surveys** and **swarming surveys.**

8.2 Bat activity surveys – manual and automated/static

8.2.1 Description and aims

Manual bat activity **transect surveys** involve ecologists walking predetermined transect routes in order to observe, listen for and record bats in flight away from their roosts using hand-held bat detectors and recorders. **Automated/static activity surveys** involve bat detectors being deployed at fixed locations to record bat activity remotely. These are usually used in combination with transect surveys.

The aim of these surveys is to answer the questions posed in Section 8.1. The results of these surveys can then be used to inform the need for further surveys or to facilitate an impact assessment and the subsequent design of appropriate mitigation.

8.2.2 Equipment

Generic documentation/equipment required for field surveys for bats is provided in Section 2.5.2; survey-specific equipment is listed in Appendix 1.

8.2.3 Expertise and licences

Section 2.5.1 discusses expertise and Section 1.2.2 provides information on licences. Activity surveys are unlikely to disturb bats if carried out correctly; however, it is good practice for these surveys to be designed and carried out, or at least led, by licensed surveyors who have gone through a period of training and evaluation.

8.2.4 Methods

8.2.4.1 Transect surveys

Appropriate transect routes should be determined during the fieldwork carried out as part of the preliminary ecological appraisal (see Chapter 4). This survey should have identified the different habitats in the survey area that will be impacted by the proposed activities and may have assessed suitability (see Section 4.3.4). All habitats should be sampled but the habitats identified as having moderate or high suitability for bats are likely to be the main focus of the transect surveys.

The extent and arrangement of the different habitats on site should inform the number and arrangement of transects required to complete the survey. This is also influenced by ease of accessibility and navigation. Some habitat types (for example, wetlands or dense scrub/woodlands) may constrain transect surveys and increase the emphasis on collecting data from spot counts, timed searches (see Section 8.2.5) or automated/static surveys (see Section 8.2.4.2).

Ideally, ecologists should have the opportunity to walk transects during the daytime in order to avoid getting lost; to identify hedge or watercourse crossing points; and to identify any particular hazards. It is more appropriate for this work to be carried out in pairs; for ecologists to know where other colleagues will be on site; and for the method of communication to be identified. This may require two-way radios in the absence of mobile phone signal.

During transect surveys, an ecologist should walk at a fairly constant speed (so the sampling area is the same per unit time) along a planned route recording observations of bats such as number of bats, flight direction, flight height, behaviour (e.g. commuting or foraging – the latter can be identified through hearing feeding buzzes), appearance and relative speed. Much of this is qualitative information that cannot be recorded using the automated systems described in Section 8.2.4.2, although obviously constrained by light levels (more so in cluttered habitats). All echolocation calls should be recorded and subsequently analysed to species or genus (see Chapter 10) even if the ecologist has attempted to identify the species by ear in the field.

Technology is available to record each bat echolocation call and link it to a specific location (using GPS points) and time to enable the data to be easily mapped and presented in reports, although some ecologists still use paper recording forms to record time, location, species and behaviour.

Because an ecologist is only in one location at a given time, it is likely that bat activity will be missed. Repeating a short transect twice during the course of one evening, randomly varying the starting point through the season and/or supplementing transect surveys with automated/static detector surveys can help to overcome this limitation. Different methods facilitate different types of analysis. For example, randomising the start point across a suite of surveys facilitates the production of a kernel density plot of the activity along the transect (see Figure A7.6).

Ideally, all habitats represented on site should be sampled by transects during a single survey visit to allow a comparison of bat activity across the site. However, if few ecologists are available and the site is particularly large it may be necessary to visit the site (covering different transects each time) over several consecutive nights (collectively considered to be 'one survey visit') to cover all areas.

Transect surveys can be undertaken as:

- **dusk surveys only** – this is likely to be the most effective method in the spring and autumn when conditions are likely to deteriorate in the night and may cause bats to go back to their roosts and not emerge for a second time before dawn;

- **dusk and pre-dawn surveys with a break between the two** – this is a useful method if the conditions are appropriate for pre-dawn activity but long nights mean a dip in bat activity is experienced in the middle of the night;

- **dusk to pre-dawn surveys** – this is most useful on short summer nights when activity levels remain high, or where the aim is to record particular types of bat activity in the middle of the night such as mating or swarming along with dusk and dawn activity;

- **pre-dawn surveys only** – these may be used to record specific pre-dawn behaviours such as bats commuting back to a roost in a particular direction.

Where multiple transects are carried out at one site, they should all be approximately the same length. A good guide is 3–5km, but transects may be shorter than this depending on the site, ground conditions, whether or not stopping points are used and levels of bat activity.

8.2.4.2 Static/automated surveys

The use of static/automated detectors facilitates quantitative analysis of the data to supplement the often qualitative data collected during transect surveys. Some examples of strategies that can be used to identify bat detector locations are given below (please refer back to Section 2.2.6 on data analysis):

- **Random:** a random sampling strategy is a good method for not introducing bias (distortion) to the subsequent analyses. The survey area is divided up into a grid (10 × 10) of equal squares and rather than surveying all squares, 20 squares are chosen randomly from the 100. Each square is numbered from 1 to 100 and 20 numbers are generated randomly, between 1 and 100, and assigned to a square.

- **Systematic:** a systematic sampling strategy is a good method for not introducing bias (distortion) to the subsequent analyses. The survey area is divided up into a grid (10 × 10) of equal squares and rather than surveying all squares, every 5th square is chosen. Each square is numbered sequentially 1 to 100 and then squares 1, 5, 10, 15, 20, etc... are selected.

- **Judgemental:** sampling locations in the survey area are chosen subjectively. For example 20 sampling locations, using the example area above, are determined based on expert opinion (after the preliminary ecological appraisal; see Chapter 4) or historical information. The approach could be described as 'haphazard' and at the extreme can fall into 'convenience' (sampling at convenient places (or times)). Judgement sampling has inherent uncertainty, cannot be readily quantified and statistical methods cannot be applied. However, this approach may facilitate the chances of recording, for example, quieter calling bats (see Table 3.7 on page 31).

- **Stratified:** the survey area is divided unequally into sub-areas allowing a sub-area(s) of interest to be surveyed more intensively (identified during the preliminary ecological appraisal; see Chapter 4). Sub-areas can be analysed individually but care should be taken when looking at the area as a whole because a bias has been introduced; some areas have been surveyed more than others. One way of looking at the whole area, while surveying sub-areas more intensively, is to pair or group sample locations by factors and use the factors in the analysis. Factors are most useful when they are simple and easily defined:
  - Field 1 – Field 2 (adjacent to Field 1 and same area)
  - Hedgerow – Watercourse (same length)
  - Woodland – Open field (same area)

Random, systematic, judgemental and stratified sampling strategies also apply to the timing of surveys; the convention in bat surveying is to use timings that are systematic.

Ideally, the same model of automated/static bat detector should be used across the site, all detectors should be deployed with the same settings and all detectors should be subject to regular testing/calibration as appropriate to avoid the introduction of bias and to allow a meaningful comparison of the results.

The microphone should be positioned to maximise the amount of bat activity recorded – this requires knowledge and consideration of the directionality/sensitivity of the particular microphone used. The choice of microphone (uni- or omnidirectional) will depend on the objectives of the survey – both types have their uses.

Automated/static detectors may be deployed at varying heights depending on site and project-specific factors. It is not usually appropriate to deploy a detector on the ground because this will decrease the survey volume around the microphone. The microphone should be located so that the recording of ambient (e.g. wind, running water, rustling vegetation) or any other source of extraneous noise (e.g. electrical signals) is minimised. It is also important to consider whether solid objects nearby (e.g. vegetation, built structures, etc.) will impede the passage of sound to the microphone, and adjust its position accordingly. It may be appropriate to elevate the detector above the height of a hedgerow to enable recording on both sides or to deploy the detector just below or above the canopy of a woodland. It may also be necessary to fence the detector or have livestock moved from a field if surveying in open pasture is necessary.

Data from automated/static systems is limited because there is no observational context. One hundred bat passes could represent one bat passing 100 times or 100 bats each passing once. Reality is likely to be somewhere between these two extremes. In cases where high levels of activity are recorded it may therefore be necessary to contextualise the results (i.e. is it one bat or 100 bats) using a manual transect or spot count survey. These methods are complementary – each performs a different function.

Table 8.1 provides a summary of the comparative benefits and limitations of transect and automated/static surveys.

Table 8.1 A summary of the comparative benefits and limitations of transect and automated/static surveys.

Survey type	Benefits	Limitations
Transect	• Bats can be counted • Bat behaviour can be observed (more limited as light falls and in cluttered habitats)	• Snapshot of time only • Ecologist is only in one location at any given time so could miss activity elsewhere • Subjectivity of ecologist can limit consistency, repeatability and quantitative analysis • Security of ecologists • Difficult in some habitat types (e.g. dense woodland or scrub or open homogenous habitats) • Labour-intensive fieldwork • Can't be used at height
Automated/static	• Can be deployed for long periods to pick up variability in bat activity in the absence of ecologists • Can be deployed in different locations simultaneously • Large amounts of data generated • More objective and therefore consistent, repeatable and allows quantitative analysis • Full auto identification is possible with some models, although caution should be exercised in choice and accuracy of software and reliance on results • Can be used very effectively for at-height surveys	• Bats cannot be counted • Bat behaviour cannot be observed • Large amounts of data generated, requiring significant storage capability • Lots of data analysis • Variability of weather over longer periods (though evens out over longer periods) • Security of detectors • Need to change memory cards and batteries

The results of the surveys should be recorded in a standard format and survey design should be iterative, each survey informed by the previous one. This is particularly important for automated surveys, where issues with a particular site or piece of equipment that would not otherwise be apparent may need to be addressed.

8.2.5 Complementary/alternative methods

Transect surveys may be supplemented by **spot counts**, where ecologists remain stationary for short, set periods of time (3–5 minutes) at locations along a transect route selected to represent the different habitats in the survey area. It may be appropriate to only sample at the spot count locations (rather than also recording along the transect) in habitats that are difficult to navigate or walk such as dense woodlands, wetlands or on steep terrain. This can make hearing, observing and counting bats easier because there is no noise from footfall and the ecologist can focus on the survey rather than navigation and safety.

Timed searches allow ecologists to move freely around the survey area for a set amount of time responding to any visual or acoustic evidence of bats by moving towards it. Timed searches can be used to standardise survey methods for bat species that are difficult to detect, or if bats are spread over a wide area that cannot easily be sampled using transects or spot counts. Timed counts provide a simple and effective means of obtaining estimates of relative bat activity in homogeneous or difficult terrain such as mountains or wetland bogs; landscapes with few features (moor, open farmland); and areas where it is difficult to walk around (e.g. in dense woodland, built-up areas, railway marshalling yards, etc.). Large sites can be subdivided into smaller areas; a random sample of these can be selected for sampling or each can be sampled on a different night. Searching for a set amount of time introduces an element of standardisation that can be repeated in subsequent surveys.

Vantage point surveys can provide information about the behaviour of early-emerging and high-flying bats such as noctule. Ecologists are located at vantage points around the site, so that all areas are covered. They then observe and listen for bats in flight while light levels allow, before and after sunset or sunrise. These surveys can provide information about numbers of bats and direction of travel, which gives an indication of the direction of the roost and the direction of early evening foraging grounds.

Transect surveys have been carried out using bikes or cars to cover more ground (or boats in aquatic habitats). However, the limitations of these methods should be recognised. Car surveys are particularly constrained because they focus the survey only on roads/tracks and the noise and lights of the cars could disturb some bat species (particularly species that avoid light). Quieter-calling species can easily be missed so these methods should not be used in isolation.

It may be necessary to capture bats using mist nets or harp traps in order to identify their species, gender and age to supplement activity survey information. The correct licence, skills and experience should be in place to carry out this activity and sensitive times of year should be avoided (such as the maternity and hibernation seasons). More information on capture and handling is provided in Chapter 9.

8.2.6 Timing

Recorded bat activity is dependent on the prevailing conditions at the time of the survey, which vary temporally (through the night, between nights, through the seasons and between years) and spatially (dependent on latitude, longitude, altitude, habitat, etc.).

Bat activity is also determined by what the bats are doing at different times of the year (although this is also dependent on prevailing conditions); the bat life cycle is given in Section 3.2.

The UK bat active period is generally considered to be between April and October inclusive, although April, September and October surveys are both weather- and location-dependent (October surveys are generally not acceptable in Scotland). Conditions may become more unsuitable in these months, particularly in more northerly latitudes, which may reduce the length of the survey season. Some useful data may be collected outside these months or weather conditions during these months may render surveys ineffective – professional judgement should be applied to determine the most effective activity survey period for a particular project.

It may be appropriate to survey for bat activity in the winter, particularly if there are hibernation roosts in, or close to, the survey area. Foraging habitats close to hibernacula may be particularly important because during the winter bats need to minimise energy used to gain food during milder weather conditions. Automated/static surveys are likely to be the most efficient way of collecting data on winter bat activity.

Table 8.2 gives recommended timings for activity surveys.

Table 8.2 Recommended start and end times for activity surveys.

Survey type	Start time	End time
Dusk survey – bat activity	Sunset[a]	2–3 hours after sunset
Dusk survey – swarming	2 hours after sunset	5 hours after sunset
Dusk to pre-dawn survey	Sunset	Sunrise or later if bats still active
Pre-dawn survey	2 hours before sunrise	Sunrise or later if bats still active
Automated bat detector survey	30 minutes before sunset	30 minutes after sunrise

[a] Adjust to earlier if in darker habitats such as woodland or if data justifies (e.g. if bats are already out by sunset on previous surveys or automated detectors show pre-sunset activity).

Timings may be adjusted (earlier or later) if necessary by the ecologist, bearing in mind the site-specific circumstances, although this should be fully justified in the survey report.

8.2.7 Survey effort

When planning surveys it is important to take a proportional approach. The number of transects, automated/static surveys and repeat visits decided upon should be proportional to the factors described in Section 2.2.5. To briefly recap, with particular reference to activity surveys, these are:

- likelihood of bats being present;
- likely species concerned;
- numbers of individuals;
- type of habitat affected;
- predicted impacts of the proposed development on bats;
- type and scale of proposed development.

An activity survey should provide a representative sample of the bat activity in all habitats present at the proposed development site (see Section 8.2.4.1). Sampling should be designed to provide a sufficient amount of data to assess the potential impacts of the development on bats.

Bat activity is inherently variable from night to night, with this variability not explained by weather conditions alone (Scott and Altringham, 2014), and so multiple consecutive nights of survey with automated systems are recommended.

Table 8.3 gives guidelines on the number of bat activity surveys recommended to achieve a reasonable survey effort on sites with low, moderate and high-quality habitat for bats (as defined during the preliminary ecological appraisal fieldwork; see Table 4.1 on page 35). Please note that the elements outlined in Section 2.2 should be considered alongside habitat suitability in designing surveys. In particular, the potential impacts of the proposals (Section 2.2.2) and proportionality (Section 2.2.5).

Table 8.3 Guidelines on the number of bat activity surveys recommended to achieve a reasonable survey effort in relation to habitat suitability.

Survey type	Low suitability habitat for bats[a]	Moderate suitability habitat for bats	High suitability habitat for bats
Transect/spot count/timed search surveys	One survey visit[b] per season (spring – April/May, summer – June/July/August, autumn – September/October)[c] in appropriate weather conditions for bats Further surveys may be required if these survey visits reveal higher levels of bat activity than predicted by habitat alone	One survey visit[b] per month (April to October)[c] in appropriate weather conditions for bats. At least one of the surveys should comprise dusk and pre-dawn (or dusk to dawn) within one 24-hour period.	Up to two survey visits[b] per month (April to October)[c] in appropriate weather conditions for bats. At least one of the surveys should comprise dusk and pre-dawn (or dusk to dawn) within one 24-hour period.
AND			
Automated/static bat detector surveys[d]	One location per transect, data to be collected on five consecutive nights per season (spring – April/May, summer – June/July/August, autumn – September/October)[c] in appropriate weather conditions for bats	Two locations per transect, data to be collected on five consecutive nights per month (April to October)[c] in appropriate weather conditions for bats	Three locations per transect, data to be collected on five consecutive nights per month (April to October)[c] in appropriate weather conditions for bats

[a] If the habitat has been classified as having low suitability for bats, an ecologist should make a professional judgement on how to proceed based on all of the evidence available. It may or may not be appropriate for bat activity surveys to be carried out in low suitability habitats. However, caution should be exercised in fringe areas (e.g. some areas of Scotland) where 'low suitability habitat for bats' may be extremely important to local bat populations due to the relative scarcity of better habitats. In such situations, bats are likely to also be more widely dispersed and may use a larger number of sites, therefore survey effort may actually need to be increased to detect use on the proposed site in question.

[b] A survey visit should aim to cover all habitats represented in the survey area that could be impacted by the proposed activities. This may consist of a single transect carried out on a single night for small sites (e.g. small housing developments) with low habitat diversity but could range up to multiple transects carried out over one or several nights (depending on number of ecologists) on a larger site (e.g. road schemes) with greater habitat diversity.

[c] April, September and October surveys are both weather- and location-dependent. Conditions may become more unsuitable in these months, particularly in Scotland, which may reduce the length of the survey season.

[d] Detector locations should be assigned to cover all habitats represented in the survey area that could be impacted by the proposed activities. This could mean a single detector location at a small site with only one habitat represented but could range up to many detector locations on larger sites. Automated/static surveys are particularly useful when assessing collision risk, e.g. detectors can be placed at crossing points on proposed roads or railways.

Note: Multiple survey visits should be separated by at least two weeks, preferably longer, to observe temporal changes in activity.

Survey data should be analysed as soon as possible and preferably before the next survey (see Section 10.2).

It is important to consider how effective the surveys are in recording species that are more difficult to detect (see Section 3.9) or exhibit highly variable or seasonal patterns of activity. It may be appropriate to adjust the survey methods, increase the number of survey nights or adjust the survey frequency to ensure these species are not under-recorded. Skalak *et al.* (2012) reported that relatively few nights are needed to detect common species but longer sampling periods may be necessary to detect rarer species. The same is true of those species that use quiet echolocation calls (see Table 3.8 on page 32).

Comparing transect and static data may also indicate that species are being recorded by one type of survey but not another, so that subsequent surveys can be adjusted accordingly.

8.2.8 Weather conditions

Please refer to Section 2.6.1 for guidance on weather.

8.2.9 Next steps

The next steps will depend on what has been recorded during the activity surveys. It may be necessary to carry out more targeted activity surveys in subsequent years or use alternative methods to gain specific information (e.g. using a trapping survey to distinguish between *Myotis* or *Plecotus* species or to define breeding status of the bats; see Chapter 9).

Where enough information has been collected, the data should be used to inform an impact assessment and the design of a mitigation strategy.

8.3 Swarming surveys – acoustic

8.3.1 Description and aims

Swarming surveys are carried out to identify if a site is used by bats for autumn swarming, which was described by Fenton (1970) as 'the flight of bats through hibernacula in late summer and early fall'. This usually occurs in the UK from August to October inclusive and activity peaks 3–4 hours after sunset (Rivers *et al.*, 2006; Glover and Altringham, 2008): observations made during the first 2–3 hours after sunset may not detect it. Autumn swarming should not be confused with what is commonly termed 'dawn swarming', where one or more bats fly around outside their roosts prior to entry at dawn.

Autumn swarming behaviour has been recorded mostly at the entrances to and outside underground sites such as caves, mines and tunnels but has also been observed around other structures such as castles, and large barns. Evidence from the Netherlands shows mass swarming events of common pipistrelle bats in the autumn followed by mass hibernation in a diverse range of building types in urban environments (Korsten *et al.*, 2015). This phenomenon requires some research in the UK but ecologists should be aware of the potential for larger numbers of this species to be present during the autumn and winter in large buildings in highly urbanised environments.

Swarming behaviour is common among *Myotis*, *Plecotus* and *Barbastella* species. Swarming probably has several important functions: mating, transfer of information about hibernation sites to young, collection of information on the condition of hibernation sites prior to hibernation and migration stopover, but as yet most lack direct evidence to support them. There is, however, good behavioural and genetic evidence to show that mating is an important function (Thomas *et al.*, 1979; Kerth *et al.*, 2003; Rivers *et al.*, 2005; Furmankiewicz and Altringham, 2007).

Rivers *et al.* (2006), in a study of four North Yorkshire caves, found that Natterer's bats undertook seasonal migration between the caves and their nursery sites over at least a 60km radius area. Between 300 and 400 bats visited the caves each night, with many more at the peak of the season. Numbers of bats vary between sites and from night to night at the same site. Activity typically starts in August and rises to a peak in September or early October before slowly declining. Many thousands of bats may visit some sites, but swarming behaviour may involve no more than a few bats each night at minor sites.

Swarming sites can therefore be important mating sites for large numbers of bats and are important for gene flow (Kerth *et al.*, 2003; Rivers *et al.*, 2005; Furmankiewicz and Altringham, 2007). Many underground swarming sites are also hibernation sites and it is likely that those bats swarming at a site go on to hibernate in the same site (Glover and Altringham, 2008). Individual bats show very high fidelity to a single swarming site (Rivers *et al.*, 2005, 2006; Glover and Altringham, 2008) and few bats are recaptured at other sites, even those close by.

The impact of destroying or changing a swarming site for development purposes is likely to be severe, so it is particularly important to investigate further whether swarming is a possibility. The aim of carrying out acoustic bat activity surveys at potential swarming sites is to establish actual use of the site by swarming bats and understand how bats use the site. If a site likely to be impacted by development does support swarming bats it is possible that further surveys will be necessary (see Chapter 9).

8.3.2 Equipment

Generic documentation/equipment required for field surveys for bats is provided in Section 2.5.2; survey-specific equipment is listed in Appendix 1.

8.3.3 Expertise and licences

Section 2.5.1 discusses expertise and Section 1.2.2 provides information on licences. Acoustic swarming surveys are unlikely to disturb bats if carried out correctly; however, it is good practice for these surveys to be designed and carried out, or at least led, by licensed surveyors who have gone through a period of training and evaluation. If disturbance to bats is likely (e.g. because an ecologist needs to deploy a detector in the entrance to a roost), then a survey licence is required.

8.3.4 Methods

The simplest and most efficient way to investigate whether bats are swarming at a site is to deploy an automated/static bat detector and recorder to record swarming bats just outside or within the entrance to an underground site (or complex structure). Repeated peaks in ultrasonic activity, reaching a maximum 3–4 hours after sunset, indicate the site is used by swarming bats and the echolocation calls recorded can be analysed to species or genus after the survey. This method is likely to generate a large amount of data at a swarming site because of the high levels of activity generally observed. However, it is unlikely to be necessary to scrutinise all recordings made, depending on the aims and objectives of the survey. Alternatively, data collected could be reduced by recording for only a few hours during the middle of the night (e.g. 2–5 hours after sunset).

8.3.5 Complementary methods

It may be appropriate to trap bats at a swarming site if it is necessary to confirm species, particularly if Annex II species such as barbastelle and Bechstein's bat may be present. Wherever and whenever possible, harp traps should be used in preference to mist nets due to the possibility of catching large numbers of bats. More information on trapping is provided in Chapter 9. With the evolution of more reliable software for automated identification from echolocation calls, trapping to determine species only may eventually become unnecessary.

Trapping to establish gender is unnecessary because the pattern of use at swarming sites is well documented: both sexes are present, but males outnumber females, consistent with mating behaviour during swarming (Thomas *et al.*, 1979; Kerth *et al.*, 2003; Parsons *et al.*, 2003a; Rivers *et al.*, 2005; Furmankiewicz and Altringham, 2007; Glover and Altringham, 2008).

8.3.6 Timing

Swarming surveys should be carried out from mid-August to October inclusive, but if only a limited survey period is available mid-September to early October is best. Species composition varies throughout the swarming season, with *Plecotus* and most *Myotis* species either peaking early or showing no discernible peak, and Natterer's bat peaking late in the season (Parsons *et al.*, 2003b; Rivers *et al.*, 2006; Glover and Altringham, 2008). Bechstein's bat, Alcathoe bat and barbastelle also swarm, but numbers caught are too low to reveal temporal patterns. Most sites have a similar mix of species: brown long-eared bat and the *Myotis* species expected in the area. It is typical for Natterer's bat to greatly outnumber all other species, particularly from mid-season onwards. Non-swarming species may also be recorded, particularly horseshoe bats, depending upon the nature of the adjacent habitat.

8.3.7 Survey effort

At least five nights of survey with an automated/static detector (in appropriate weather conditions for bats; see Section 2.6.1) in each month of the swarming season of mid-August to the end of October is recommended to establish whether a site is used for swarming or not.

If trapping is undertaken, then recommendations on survey effort are provided in Chapter 9.

8.3.8 Weather conditions

Please refer to Section 2.6.1 for guidance on weather.

Many studies have noted that bat activity at swarming sites varies markedly from night to night: bat activity is significantly suppressed by rainfall and positively correlated with residual maximum ambient temperature. Grubb (2012) also found high winds depressed activity. Moon phase does not appear to influence swarming activity (Parsons *et al.*, 2003a), but a bright moon has been known to lower capture success (if trapping) at exposed locations. Swarming activity appears to be more likely when weather conditions are more stable so targeting periods of high pressure may be appropriate.

8.3.9 Next steps

See Chapter 9 regarding trapping bats at swarming sites. This is only likely to be necessary if Annex II species may be present. If the presence of Annex II species is unlikely, then trapping is less appropriate because species assemblages using swarming sites are well documented from other studies (Parsons *et al.*, 2003b; Rivers *et al.*, 2006; Glover and Altringham, 2008).

Swarming sites are also used for hibernation so it may be necessary to also carry out hibernation surveys as described in Section 5.3.

8.4 Back-tracking surveys

8.4.1 Description and aims

Back-tracking surveys involve ecologists making visual observations of bats commuting away from their roosts at sunset or commuting back to their roosts at sunrise then attempting to track back to the roost based on these observations. Bat detectors are also used to record echolocation for identification of species, where possible. This technique was first developed in the Netherlands and is based on four principles:

- The earlier a bat is seen after sunset or the later it is seen before sunrise, the closer it is likely to be to its roost (the exact time depends on the species).
- Bats fly away from their roost at sunset, so ecologists should move in the opposite direction as the bats at this time to locate the roost.
- Bats fly towards their roost at sunrise, so ecologists should move in the same direction as the bats at this time to locate the roost.
- At sunrise, some bats species swarm at roost access points for between 10 and 90 minutes before entering.

The aim is to find roosts by making observations of commuting bats. These surveys are often used after a bat activity survey if numbers of bats were seen all commuting in one direction and follow-up is required or in situations with lots of potential roosts sites that are difficult to survey using alternative methods (e.g. in woodlands or highly urbanised areas).

8.4.2 Equipment

Generic documentation/equipment required for field surveys for bats is provided in Section 2.5.2; survey-specific equipment is listed in Appendix 1.

8.4.3 Expertise and licences

Section 2.5.1 discusses expertise and Section 1.2.2 provides information on licences. Back-tracking surveys are unlikely to disturb bats if carried out correctly; however, it is good practice for these surveys to be designed and carried out, or at least led, by licensed surveyors who have gone through a period of training and evaluation. If disturbance to bats is likely (e.g, because an ecologist would needs to investigate the roost when found using a torch or endoscope), then a survey licence would be required.

8.4.4 Methods

Ecologists should be deployed on potential or actual commuting routes close to roost sources (identified during the preliminary ecological appraisal, see Chapter 4, or during activity surveys, see Section 8.2) and note the time and direction of travel of each bat encountered on a detailed plan of the site. The ecologists should move in the opposite direction to the bats at sunset and in the same direction as the bats at sunrise. As ecologists approach potential roosts they should watch for bats emerging or dawn swarming at roosts.

If multiple ecologists are used they should be in constant contact via hand-held radio to communicate their observations. The data from multiple ecologists can also be pooled for a bigger picture of bat activity across the site, which can be used to design subsequent surveys where necessary.

In theory, back-tracking surveys work best for species with loud echolocation calls which form large roosts, but they can potentially be used to locate the roosts of any bat species.

8.4.5 Complementary methods

Back-tracking surveys are rarely used in isolation; they are most effective when combined with roost (Chapters 5 to 7) and bat activity surveys (Section 8.2).

8.4.6 Timing

As back-tracking surveys are most effective for larger roosts, the best time to carry them out is between May and August, when maternity colonies are gathered. However, results may be gained if carried out in April, September or October, depending on the individual situation (although October surveys are not considered appropriate in Scotland). See Section 8.2.6 for further comments on timing of activity surveys through the year. Table 8.4 gives recommended timings for back-tracking surveys during the night.

Table 8.4 Recommended start and end times for back-tracking surveys.

Survey type	Start time	End time
Back-tracking survey at dusk	15 minutes before sunset	When it is too dark to observe bats or when the source roost has been found
Back-tracking survey at dawn	2 hours before sunrise	When bats cease to be active or when the source roost has been found

Timings may be adjusted (earlier or later) if necessary by the ecologist, bearing in mind the site-specific circumstances, although this should be justified in the survey report.

8.4.7 Survey effort

The survey effort for back-tracking surveys is not fixed. These surveys have the specific aim of locating roosts using commuting bats for guidance and should be continued until this aim is reached unless alternative methods are considered more appropriate.

8.4.8 Weather conditions

Please refer to Section 2.6.1 for guidance on weather.

8.4.9 Next steps

If a roost is found during a back-tracking survey it may be necessary to follow up with a roost characterisation survey (see Section 7.2) to count the numbers of bats present at the roost.

Chapter 9

Advanced licence bat survey techniques

9.1 Introduction

Being small, nocturnal and with many species being morphologically and acoustically similar, bats remain one of the most challenging groups of species to study for the purposes of determining impacts from development, especially when working to the deadlines often associated with a commercial project. While research on the ecology of some bat species is widely available, there are still significant gaps in the knowledge about the basic ecological requirements of many species. Radio tagging and tracking surveys are therefore powerful survey tools to obtain information on bats and bat populations potentially affected by a proposed development. However, radio tagging and tracking surveys do involve significant levels of risk to bats, and therefore these guidelines have been written to take account of Eurobats Resolution 4.6, which provides guidance on the capture and study of captured wild bats.[45]

This chapter provides guidelines on using ALBST and principally concerns the trapping of free-flying bats and, where required and appropriate, the subsequent attachment of radio transmitters. The techniques covered in this chapter need to be specifically licensed by the relevant licensing authority.

Deciding to use ALBST is a process of balancing the data requirements to meet the objectives of the survey with the level of potential impact on bats or bat populations from using the technique. The decision-making processes should also fully consider the potential level of impacts from the proposed development (see Section 2.2). More detailed information gained from ALBST is likely on projects with greater impacts on 'difficult to survey' bat species such as tree-roosting or quiet-calling species; more sensitive bat populations, such as Annex II bat species generally; SACs or SSSIs designated for bats; or in particular habitats such as woodland. However, it should be recognised that using such techniques also poses a risk to sensitive bat populations.

A point of principle is that where the required information can be obtained using non-invasive techniques, these should be used first. However, while non-invasive methods of surveying bats such as bat activity surveys have dramatically improved data gathering for development-related projects, such techniques have limitations. In particular, confidence in identifying bat species such as *Myotis* bats (unless species-specific behaviour has been observed, as is the case with Daubenton's bats flying close to the surface of water) is extremely difficult (Parsons and Jones, 2000; Walters *et al*., 2012). In addition, quiet echolocating species (or those that do not call while foraging) often go under-recorded and non-invasive survey methods are generally unable to confirm the sex, age class or breeding status of individual bats, especially away from the roost.

If the potential impact of development activities is unlikely to significantly affect bats or their habitats, this should be reflected in the survey design and the use of ALBST is unlikely to be necessary. Equally, projects or developments (of any scale, from small barn conversions through to major road schemes) that are likely to have high direct or indirect impacts on bats (particularly for rarer or uncommon species or at the landscape level where impacts may affect multiple bat species and populations) will be required to have much more detailed data sets, potentially justifying the use of ALBST. Box 3 provides an example of the effective use of ALBST.

Box 3 Example of effective use of ALBST.

A series of trapping and simultaneous full-spectrum bat detector surveys were undertaken in the same woodland habitat over six months during the bat active period of 2014. In total, 82 bats were captured and approximately 3500 bat recordings were made over 17 survey nights. Only six bat detector recordings could be assigned to long-eared bats whereas 41% of the bat captures were of brown long-eared bat. Furthermore, three Bechstein's bats, two of which were from a nearby newly discovered breeding population, were captured. These results highlight the significant under-recording of species that listen rather than echolocate and where trapping is often the most effective tool to confirm their presence. Given the scale of the housing development proposals (over 5000 units), the potential impact on the woodland from the development (lighting and increased recreational use), as well as the possible presence of rare species in the general area, the use of ALBST was appropriate and **provided information to inform the EIA that other techniques could not achieve.**

[45] Found at http://www.eurobats.org/sites/default/files/documents/pdf/Meeting_of_Parties/MoP4_Res.6_ Issue_of_Permits.pdf) and states that 'radio-telemetry should only be used for well-organised and authorised projects where essential data cannot be acquired with less-intrusive methods'.

Radiotelemetry can provide valuable data on roost use, activity patterns, colony and individual home ranges, foraging behaviour and habitat use. For impact assessments associated with development, this data can provide useful context on how important a proposed site might be within a bat population's home range and whether preferred foraging, commuting or roosting habitat types will be affected, enabling the design of more effective mitigation. Furthermore, radiotelemetry can locate roosts of challenging species (especially in trees).

It is important to highlight that radio tracking surveys are essentially population sampling methods. It is never necessary or desirable from a bat welfare perspective to mark every animal from a population, and only sufficient bats to confidently represent the population being investigated should be tracked. However, this approach can be misrepresented in development projects as the focus for impact assessments and/or mitigation is often on only the individual bats being tracked and their movements, rather than using the sampling to identify which **type** of commuting routes or foraging habitats the population is likely to use. This issue is best overcome by proper study design and statistical testing of the samples used. All effort should be made to extract as much information as possible from a marked individual to justify the method. It is not considered acceptable given the intrusive nature of the methods on bats and the costs of such surveys, for any subsequent analysis to be limited to simple dots on a map, unless roost location is the only objective. More information is provided in Section 10.4.

As highlighted earlier, this technique should only be used in cases where other options for obtaining data are ineffective or grossly inefficient and the level of potential impact on important bat populations is considered high, such as the loss of significant high-quality bat foraging or roosting habitat. For instance:

- High-impact developments at a landscape scale that may affect substantial roosting and foraging areas for a wide assemblage of bat species, especially those difficult to identify through bat detector systems.
- High-impact developments at a landscape scale affecting rare bat species, for example, Annex II species or features of SSSIs.
- High-impact developments on areas likely to support proportionately higher populations of tree-roosting bats or bats likely to be in inaccessible roost types (quarry faces, etc.), where other methods have not been able to locate roosts likely to be present.

Although these guidelines are focused on single-site/project-related developments, radio tracking of key populations can also be effectively used to provide a strategic approach to land use/development-related planning, particularly around sites supporting Annex II species. For instance radio tracking can be used to identify key habitats and sustenance zones around bat SACs to inform HRAs and local development plans. This is likely to be more efficient and productive than undertaking a site-by-site approach to gathering such information.

These guidelines should be read in conjunction with NE's advice regarding the use of these techniques (WML-G39 2013, NE, 2013).

9.2 Trapping surveys

9.2.1 Description and aims

This section focuses on the capture of free-flying bats with mist nets and harp traps. This technique can be used at bat roosts, bat swarming sites and bat commuting and foraging areas.

Given its rarity, quiet echolocation calls and the difficulty of reliably separating *Myotis* bats from echolocation calls (Parsons and Jones, 2000; Walters *et al.*, 2012), species-specific guidelines are given for surveying Bechstein's bats where developments are likely to affect this species and/or its habitats.

The need to undertake trapping surveys will depend on a range of factors and, in particular, the questions requiring answers to inform an impact assessment. Recommended use of these techniques include:

- To determine species identification: for instance if bat detector surveys have found proportionately high levels of *Myotis* bat activity and the development is likely to have a high impact on the habitats of such species, then it will be important to confirm which *Myotis* species are present to inform the impact assessment and mitigation strategy. It is also essential to identify bats to species level for high-impact licensing purposes when other techniques have been unable to do so.
- To determine gender and breeding status: trapping can be used to determine gender and breeding status and is particularly important when the impacts of a development on a roost or site are high (i.e. full destruction) and knowing the breeding status of a population is crucial to designing the most appropriate mitigation. In addition, understanding the breeding status of bats using foraging or other non-roost sites can be an important element of valuing the importance of the site for impact assessment purposes.
- To gain further information about rare or under-recorded bats: the presence, gender, breeding status, roost locations, foraging areas and commuting routes of rare species such as grey long-eared bat, barbastelle and Bechstein's bat may need to be confirmed where they could be present and when their potential habitat is affected by the proposed development.
- To find tree and building roosts at a landscape level: if high impacts on bats are anticipated, then trapping can be used to determine the presence of breeding bats and the selection of such individuals for the attachment of radio transmitters. This is an effective approach to locate breeding colonies, particularly tree roosts.

It should be noted that trapping surveys also have their own biases and limitations and may be more effective at determining the presence of certain species (for instance those species generally found in cluttered habitats). Data collected using this technique should be considered alongside other techniques to provide a balanced data set of bats using any particular site.

9.2.2 Equipment

Generic documentation/equipment required for field surveys for bats is provided in Section 2.5.2; survey-specific equipment is listed in Appendix 1 and further information about mist nets, harp traps and lures is provided in Appendix 5.

9.2.3 Expertise and licences

Section 2.5.1 discusses expertise and Section 1.2.2 provides information on licences.

These techniques can significantly affect the welfare of bats and therefore bat handling and identification skills need to be regularly practised to be able to extract and process bats quickly. Experience of handling wild bats from a range of species including small, medium and large bats should be kept up to date.

If acoustic lures are used (see Section 9.2.5), continual training or experience in the most effective use of lures is recommended, because of their evolving nature.

Licences from the relevant licensing authority (see Section 1.2.2) are required to use these techniques, including the use of lures. Using lures without traps to attract bats also requires a licence.

9.2.4 Methods

The first stage of a trapping survey is the identification of potential trapping sites through a review of site plans, aerial imagery, the proposed activities and any existing habitat/bat-related data. This information helps to identify the sites that would increase the likelihood of catching bats in relation to those areas impacted by the proposed development. This should be followed by a daytime site visit to determine the micro-siting of the traps. Large projects involving multiple trapping sessions will require the relevant project licence and broad trapping locations and methods will be required as part of any project proposal when applying for a licence.

Trapping using mist nets and harp traps can be passive (without lures) or active (with lures used to attract bats; see Section 9.2.5). The set-up and location of traps and nets will vary depending on which method is being used and whether specific species are being targeted. Passive trapping with mist nets and harp traps should be based on the principle of pinch points or funnelling the bats to the traps. Mist nets have the advantage of covering more space and being lighter, but they require continuous monitoring and higher levels of bat handling skills to extract bats. Bats are more easily extracted from harp traps; however, they cover less trapping area and are heavy.

Recommended trapping locations include areas where vegetation or other structures limit the space through which bats can fly or manoeuvre, therefore increasing the chance that the bat will fly through the restricted space where the net or trap is located, for instance:

- woodland rides and edges with overhanging tree branches;
- streams/river corridors and bridges;
- low-hanging branches from large isolated trees;
- gaps in treelines/hedgerows;
- next to water features such as lakes/rivers, especially adjacent to riparian woodland;
- tunnel, cave and mine entrances and passages; or
- barn doors.

Placing traps next to building features such as hanging tiles are also effective when trapping at building roosts or swarming sites (see Mitchell-Jones and McLeish, 2004, for more detailed information on such techniques).

Where larger numbers of bats are expected, harp traps are likely to be safer than mist nets because of the need to extract bats from mist nets soon after they have flown in.

It is essential to ensure when working in or around water that bats will not be drowned if they become trapped and their weight drags the net or capture bag into the water.

Mist nets should be monitored continuously when deployed, ideally using night-vision equipment or, as a minimum, a bat detector to monitor any bat activity around the net. Following any bat detector activity (or at 5-minute intervals) nets should be checked to ensure captured, quiet-echolocating bats do not remain unnoticed. Nets should be checked with a powerful torch (ideally with a red filter to preserve night vision) very quickly to avoid putting off bats that may otherwise fly into the net. Harp traps should be checked ideally every 15 minutes.

Where a bat is caught it should be extracted from the mist net or harp trap as soon as possible and released by the ecologist after obtaining the minimum information (which should be labelled with date and time of capture and trapping location) as follows:

- species;
- sex;
- age class (where possible);
- breeding status (pregnant/lactating/post-lactating/non-breeding).

Processing bats should be carried out as quickly as possible to obtain the data required. Ideally bats should be handled as little as possible and released nearby within minutes of being captured. This is especially important for breeding females or during the cooler active months such as September and October. Non-target species, stressed or heavily pregnant bats should be released immediately with no processing. If heavily pregnant bats are being caught unexpectedly then consideration should be given to ceasing trapping entirely.

Forearm or other morphological measurements are generally used to help identify the bat species, therefore prolonged handling for these purposes should only be undertaken where identification is proving challenging. If species identification can be made without taking such measurements, then this part of the process is generally superfluous. Furthermore, for commercial surveys, weight data is only of use where bats are to be marked with radio transmitters, therefore it is unnecessary to weigh bats for trapping purposes as this adds unnecessary time to the processing, potentially creating problems for release.

While the bat is under the control of the ecologist, it is important to ensure the equipment used to hold the bat(s) and the processing stages comply with licensing conditions and guidance (e.g. NE, 2013).

Bats should be released at height and for most species releasing at head height is sufficient. Noctule bats may struggle to launch at this height and it is often necessary to find a suitable tree and allow this species to climb to a height where it is comfortable to launch. When releasing bats it is important to continually monitor behaviour to identify whether bats are fit to release and have launched successfully.

9.2.5 Complementary methods

Bat activity surveys (see Chapter 8) are complementary to trapping and can provide a more balanced data set than trapping alone, subject to the objectives of trapping. Care should be taken to ensure that acoustic surveys do not record calls emitted by the lures (see below) when trapping and acoustic surveys are undertaken simultaneously in the same locations.

Where sites are located within the known distribution of Bechstein's bat and suitable habitat for this species is likely to be impacted (see Bat Conservation Trust, 2013) then species-specific surveys are likely to be required. Mist nets and/or harp traps used with a lure emitting Bechstein's bat social calls is the recommended method of surveying for Bechstein's bats as these bats use quiet echolocation and even when detected using bat detectors they are very difficult to distinguish from other *Myotis* bat species (Parsons and Jones, 2000; Walters *et al.*, 2012). The use of a lure constitutes active trapping and, for this species, traps and lures should be placed in the cluttered interior area of woodland. This technique has been used to great effect with Bechstein's bats (Hill and Greenaway, 2005; Davidson-Watts, 2008; Miller, 2012).

Acoustic lures should be placed close to the net or harp trap. For harp traps, the most effective technique appears to be placing the speaker just above the catch bag in the centre of the trap as bats are more likely to be caught by the lower parts of the strings of the trap and have less time to escape. Net configurations vary and so the positioning of the lure will also vary. However, placing the lure or speaker close to the mist net will increase the chance of a Bechstein's bat being captured as it investigates the lure. Some do's and don'ts relating to lures are as follows:

- Do place lures and/or lure speakers close to the trap or net, as this increases the chance of the Bechstein's bat being captured when investigating the lure.
- Do move lures between traps and nets where there are more traps/nets than lures as this is more effective than having a stationary lure, which bats may become accustomed to. This also provides greater coverage of a site.
- Do play recognised, tested and effective Bechstein's bat social calls.
- Do have periods of silence to determine whether bat activity is present around the nets when not using the lure.
- Do consider turning the lure off during extraction to avoid unnecessary stress to the bat, particularly when extracting bats from mist nets.
- Do not use high volumes as abnormally loud calls could be counterproductive by deterring Bechstein's bats, particularly those using cluttered habitats.
- Do not use bat distress calls because the meaning of distress calls to bats is poorly understood and has the potential to have negative consequences for local populations.
- Do not use lures within 50m of known active roosts, including near bat boxes that may contain a roost, as this may cause prolonged disturbance to bats present at the time.
- Do not use lures within 100m of swarming sites during late summer/autumn as this may cause prolonged disturbance to bats present at the time.

Some precautionary advice on the use of lures is provided in Box 4.

Box 4 Precautionary note about the use of lures to aid the capture of bats in traps and nets.

> Although lures have been in use by various bat researchers and bat workers since the late 1990s, very little is known about the full effects these devices have on local bat populations. They have been shown to be very effective at increasing capture rates with harp traps and mist nets, particularly in more cluttered habitats such as woodlands and with certain species. However, no significant research has been undertaken to consider whether there are any detrimental effects of using them so they should be used with caution when all other methods have been considered and only with very specific aims and objectives.

More information on acoustic lures is provided in Appendix 5.

9.2.6 Timing

Subject to environmental conditions, trapping surveys for development-related projects should normally be undertaken between May and October when bats are most likely to be active (but not in the potentially vulnerable post-hibernation period of April unless there is a specific requirement approved under a project licence). The exact timing of the surveys will largely depend on the objectives and the potential bat habitat of the site affected. For instance, the most appropriate time to survey a potential swarming site would be between August and October, whereas trapping to confirm the presence of breeding bats should be undertaken between May and August. Unless clearly justified through the project aims, it is recommended that trapping during the period of June to mid-July is not carried out to reduce the risk of unnecessarily catching heavily pregnant bats or bats with dependent young. Trapping during this period is best covered by the relevant project licence rather than class licence.

NE Class licences (Level three for mist netting and Level four for harp trapping; see Section 1.2.2.) allow for a maximum of three trapping nights per site for commissioned developments without a specific project licence (this is not the case for the other UK countries, where the relevant project licence is required). Therefore when these techniques are used as a complementary method to other survey techniques (i.e. not under the relevant project-specific licence), such as bat activity surveys to identify *Myotis* species or surveys for under-recorded species such as Bechstein's bat, it is recommended that at least three trapping surveys are undertaken. These surveys should be spaced across the bat active season with one survey in May, a second survey in July/August and the third survey in August/September in suitable weather conditions. Trapping the same trap site locations more than once a month would require some justification from a disturbance perspective. Should more trapping sessions be required to meet specific objectives, then a project licence would be required.

On the day of the trapping survey, ecologists would normally need to arrive at the proposed trapping site(s) at least an hour before sunset to confirm exact trapping points, identify any additional health and safety issues, and set the traps. A trapping survey would usually commence at dusk and continue until 2–3am depending on conditions, capture success, general bat activity and the objectives of the survey. For instance if the objective was to capture a specific bat species for radiotelemetry, then trapping would cease once the target bat or bats have been captured. When trapping for swarming surveys, activity is likely to peak much later in the night (see Section 8.3.1) and therefore survey timings should be adjusted accordingly.

9.2.7 Survey effort

Survey effort depends on a number of factors including the size of the site, the type and quality of habitats present and the objectives of the survey. For instance surveys to trap specific species for radiotelemetry will require an assessment of suitable habitat both on and off the site, a review of previous records and an appraisal of suitable trapping areas to determine the effort required to meet the objective.

The number of harp traps/mist nets that are deployed simultaneously will depend on the extent of habitat to be surveyed. Traps/nets should ideally be no less than 100m apart when using lures.

For smaller projects where impacts are more confined to specific areas of high-quality habitat, at least three trapping surveys should be undertaken over the active period, in line with other bat activity/survey methods during late spring, summer and autumn (see Section 8.2.7), with priority areas being woodlands, treelines and wetland areas. The number of traps/nets will vary depending on the size of the areas being surveyed and the species likely to be encountered.

Large infrastructure schemes involving impacts on high-quality bat habitat such as deciduous woodlands, treelines and wetlands, with multiple trapping objectives such as the confirmation of breeding bats and the determination of bat assemblages, are likely to require many trapping nights with multiple harp traps and/or nets being used simultaneously over a 5- or 6-month period during the active bat season, especially if rarer (e.g. Annex II) or significant levels of tree-roosting species are predicted to be present. In some situations, trapping surveys over consecutive years may be relevant.

See Box 5 for more information on survey effort for Bechstein's bat.

Box 5 Survey effort for Bechstein's bat using traps and lures.

To determine the presence/likely absence of Bechstein's bat on a site, the lure and net/harp trap method should be used and trapping surveys conducted for a *minimum* of six trap nights over the active bat season. One trap night is one lure and net or harp trap combination on one night. Therefore, six trap nights can be achieved by three nights of trapping with two sets of trap/lure combination. Ultimately the total number of traps/nights will depend on the size and nature of the potential Bechstein's bat habitat available. If the site is large with multiple woodland copses or treelines with potential for this species, then more trap nights are likely to be required.

Trapping surveys for Bechstein's bats should be undertaken across the active bat season to ensure that the key stages of the breeding cycle are covered, with ideally one survey pre-parturition and one survey post-parturition between May and August, at least one month apart.

9.2.8 Weather conditions

Please refer to Section 2.6.1 for guidance on weather.

Effective trapping is subject to environmental conditions as traps are generally less effective in wet and windy conditions. This is more relevant to mist nets than harp traps, where water droplets and wind can make nets more visible to bats. In addition, trapping bats in cool and wet conditions can seriously affect their welfare, because they may go torpid in harp traps, making effective release more difficult.

Weather forecasts should always be consulted before a survey is carried out, to identify whether conditions will be favourable for trapping. **Trapping should be avoided during periods of prolonged rain (more than isolated showers, where trapping can be briefly suspended), and trapping should not be undertaken in temperatures below 8°C**, unless duly authorised by a project licence, because bats are likely to be much harder to release effectively (in any case, activity levels would most likely be low and the data produced would be constrained).

9.2.9 Next steps

Trapping is usually one of the last techniques to be used to obtain data about bats using a site, and should provide a great deal of useful information to properly inform an impact assessment. However, should the presence of rare species be confirmed and/or the results suggest that more information on tree-roosting bats is required, then the next step may be radiotelemetry (see Section 9.3), or more focused activity such as roost surveys (see Chapters 5 and 6).

Some bats such as whiskered, Brandt's bat and alcathoe bat are very difficult to identify in the hand and photographs may need to be taken for further analysis. In addition droppings from these bats (when left in clean holding bags) can be collected and sent for species identification via DNA analysis (see Appendix 4). Various universities and private companies offer this service.

9.3 Radio tagging/telemetry surveys

9.3.1 Description and aims

The aim is to effectively **mark a target bat with a radio transmitter for radiotelemetry** to obtain location data and determine the following:

- Location of roost sites
- Population and individual home ranges and core areas
- Habitat use
- Activity patterns and distances travelled

When properly analysed, location data obtained through radiotelemetry should be able to help identify how the proposed development site relates to the bat population's home range, core foraging/flying or commuting habitats and roost sites (see Section 10.4), thus enabling an effective impact assessment and, where necessary, a mitigation strategy to be developed.

9.3.2 Equipment

Generic documentation/equipment required for field surveys for bats is provided in Section 2.5.2; survey-specific equipment is listed in Appendix 1 and more information about radio tags, receivers and antennae is provided in Appendix 6.

9.3.3 Expertise and licences

Section 2.5.1 discusses expertise and Section 1.2.2 provides information on licences.

There are a number of different skills sets involved in radio tagging bats:

- **Survey design and scope** – to design an effective radio tracking survey, ecologists require a full understanding of the ecology of the bat species concerned and have experience of the practical application of these techniques, as well as data collection and analysis methods to obtain the appropriate information to inform the survey objective. No licence is required to undertake this task/role; however, it is unlikely that a suitable scope of works can be developed by ecologists without sufficient experience in using these techniques on the ground.
- **Tagging bats** – these techniques can significantly affect the welfare of bats and therefore ecologists undertaking this task require very good and regularly practised handling skills to be able to process bats and affix transmitters quickly and effectively. This task is subject to licensing from the relevant licensing authority.
- **Radiotelemetry** – a basic understanding of the physics of radio waves (when tagging with radio transmitters) is required as ecologists need to understand the limitations of this technique and how signals from transmitters are manipulated by the environment. Ecologists will also require excellent map reading, compass and navigation skills to be able to plot bat locations and take accurate compass bearings at night.

9.3.4 Methods

A significant amount of useful information on radiotelemetry design, field tracking and analysis techniques can be found in Kenward (2001). Welfare issues are covered in some detail by NE's guidance note WML-G39 (NE, 2013).

Highlighted below are the key steps and considerations that are important for bat-related tagging and tracking for development-related projects in the UK.

- **Survey design** – this stage is crucial and should be undertaken before the bat active season. Survey design will depend on the objectives of the survey. For instance, the approximate number of bats to be marked will need to be calculated/estimated. Sampling size is one of the most important factors in designing a radio tracking survey; resources should be prioritised to track more bats for less time rather than fewer bats for more time. For surveys to determine habitat use, more bats (the sampling points) than habitat categories are required to be able to use compositional analysis (a common statistical method for robust habitat preference of radio-tracked animals; see Section 10.4). This is likely to be more than five bats and may be more depending on colony size and could involve multiple species, depending on the scale and impacts of the project. There are likely to be differences in behaviour between breeding and non-breeding bats, and between different sexes and age classes (adults/juveniles). It will therefore be important to clearly identify the target bats and the reasons these are being sampled.

- **Landowner access (for off-site tracking)** – this needs to be arranged and, if this becomes a major limitation to data collection, a plan of how data will be collected from roads or other public areas (although rights of way comprise a right to cross the land, not to undertake any other activity such as survey).

- **Resource planning and licensing** – appropriate resources will need to be allocated in terms of equipment, such as tags and receivers and tracking teams. Tags and equipment will need to be ordered from suppliers with plenty of notice. It may be appropriate to check licensing turnaround times to give more confidence in timescales, particularly for bigger projects where the manpower and associated logistics need to be booked well in advance.

- **Tagging bats** – when a target bat is captured either in the roost or the wider countryside, it should be weighed initially to both ensure it is a good weight for that species and that it meets the weight requirements for tagging. Radio transmitters should be no more that 5% of total body weight. The bat should be checked over to evaluate whether it appears healthy, in good condition and is free from injury or damage. Species, age, sex and breeding status should be noted. Tagging mothers with dependent young within the roost is not recommended. All UK bats are marked by fixing the transmitter dorsally between the shoulder blades with the antenna trailing behind the bat. Fixing with suitable glue involves carefully parting or trimming the fur and applying glue to the fixing location on the bat and glue to the transmitter before attaching the tag. It can take between 10 and 30 minutes for the glue to cure sufficiently before releasing a bat. Bats should not be held for more an hour. Bats should be released (see Section 9.2.4) and post-release observations made for up to an hour to ensure the bat can fly freely and is not grounded. This observation cannot be made if bats are released back into their roosts and therefore this is not recommended. If a bat cannot fly properly following tagging, the tag must be removed if possible (by cutting the fur of the bat); the aerial should be cut off the tag; and/or advice or assistance sought from a vet.

- **Radiotelemetry** – the most basic form of data required from radiotelemetry surveys is the bat identification number, its location and the date/time the location record was made.

There are two main methods for determining a bat's location using radiotelemetry. The close-approach method involves at least one ecologist with receiving equipment following an individual bat and when the ecologist considers it has reached the bat's location, a record of the time and usually 8-figure grid reference is made. In addition, this method can make observations of behaviour and the use of habitat if close contact with the bat is maintained. This is the most accurate method of pinpointing a bat's location if the bat is relatively static, but is also constrained by land access. A significant amount of time can be spent approaching the bat before it suddenly moves quickly to another area without its position being confirmed.

The other method is triangulation, and involves a minimum of two ecologists in different locations taking simultaneous bearings at regular intervals (usually between 5 and 15 minutes) from the direction of the bat's strongest signal. This method is good for tracking multiple bats over a small area and where access to land is not possible. The accuracy of this method depends on how close the two ecologists are to the bat and their position in relation to each other and the bat. If the ecologists are closer to the bat and the lines of strongest signal are perpendicular this will increase accuracy. Triangulating moving bats at distances over 500m can achieve no more than assignment of a 6-figure grid reference. A useful method of determining the accuracy of triangulation of tagged bats in a particular study area is to use an ecologist with a tag to act as a simulated bat, from which the accuracy of bearings and triangulation fixes can be assessed under controlled conditions.

In summary, it is advisable to use a combination of both triangulation and close approach to get the most accurate data set and maintain contact with a bat. The most effective method is for three tracking teams to be deployed, with two teams triangulating a bat's broad position and the third team pinpointing the exact location using close approach.

It should be noted that while both methods are effective at obtaining location data, it is not always a reliable method of obtaining behavioural data, in that a tracked bat may be flying in a particular location, but whether the bat is foraging or socialising can be difficult to determine.

Maintaining contact with the bat is the highest priority and, with some long-ranging and fast-flying species, this is a particularly challenging task. Where contact is lost, then searching further areas in the direction the bat was last detected and in particular using high ground will increase the probability of relocating the bat. However, it should be borne in mind that, for the majority of commercial/development-related projects, tracking must at least be able to determine when the bat is using, or not using, the proposed development site.

Some species of bat (especially tree-roosting species in closed canopy woodlands) are also known to move short distances between tree roosts during the day. Therefore it should not be assumed that the equipment is faulty if the bat appears not to be in the roost it was last located in at sunrise.

Statistical analysis of radiotelemetry data to answer questions such as 'which habitats the population prefers' and 'how much time the sample bats spend on the proposed development site', or 'what proportion of home range or core flying/foraging areas are on the proposed development site' should be a major consideration to do justice to the data obtained using these methods. Further information on these techniques is given in Section 10.4.

9.3.5 Complementary methods

Bat activity surveys (see Chapter 8) in foraging areas identified through radiotelemetry are a useful complementary method where resources are available, as radiotelemetry of a small number of bats does not provide a full picture of bat activity.

Roost inspection surveys (see Chapters 5 and 6) and emergence/re-entry counts (see Chapter 7) are essential to understand the population size and therefore the appropriate number of bats to mark for radiotelemetry to meet the survey objectives. Depending on the circumstances, it might be possible to undertake a population count first and then decide on the number of bats to be marked (usually for obvious and relatively permanent roosts); however, in many situations it is likely that a target bat will be captured while foraging, enabling the roost to be found and a count subsequently carried out. This count would then contribute to the decision-making process about how many more bats to tag.

9.3.6 Timing

For consultancy purposes, radio tagging and subsequent radiotelemetry would usually take place during the active bat season unless specific objectives for winter foraging information are required. Trapping surveys are usually carried out between May and October, as discussed in Section 9.2.6. However, trapping early or late in the active season will be constrained by environmental conditions.

The specific dates of tagging and tracking bats depends wholly on the objectives of the survey. For instance, to locate maternity roosts it is advisable to undertake tracking in May, June, July or August (subject to welfare considerations). Bats have either dispersed or are dispersing from maternity roosts by September and therefore reliable population counts are unlikely.

Tagging bats will generally be linked to trapping surveys, either at the roost or in the wider countryside. It is recommended that marked bat(s) are followed immediately after tagging to gauge behaviour (and to be confident the bat is moving around). If the bat's roost is unknown, it is also advisable to stay in contact with the bat to get a likely direction of the roost as it may return there. If possible, captured bats should be followed until dawn when they return to their roost, as some bats are harder to find once inside. It is recommended that bats are tracked from roost emergence until final return. Sometimes bats will return to their roost during the night and may not re-emerge for the rest of the night. At other times bats will make numerous flying bouts from the roost and use other roosts during the night, all of which can be essential data. Additionally, bats have been recorded having separate foraging areas used at different times of the night, an early and late night foraging territory, and so it is important for bats to be continually monitored during the period of time they would be expected to be active and away from the roost.

Tagging of heavily pregnant and early lactating bats should only be undertaken where there is an overriding reason, e.g. where it fits within a detailed sequential study of a bat species through the breeding season. For roost finding, tagging should avoid such bats. When using these techniques, bat welfare should always be the overriding priority.

9.3.7 Survey effort

Radio tagging and tracking surveys should be proportionate to meet the survey objectives. The tracking of one or two bats to determine habitat use and population home ranges will not be sufficiently robust. Equally, tracking more than two bats simultaneously from the same population may be unnecessary should the objective of tagging and tracking be to locate a sample of breeding roosts (although this is species-dependent).

For surveys investigating habitat use and activity patterns of breeding colonies, at least 5–10% of the (estimated) population should be marked, and for rare species up to 25% of the animals of a population if potential impacts are high. Ultimately expert opinion, the questions of the study and statistical analysis requirements should be considered to ensure the appropriate number of animals are tracked to meet the aims of the project, and balanced against the welfare of individual bats and effects on the population. Tagging more than five bats from the same roost simultaneously should be avoided (due to the risk of entanglement) and, to this end, consideration of obtaining data over the entire season and even over two seasons is required. This is especially important for detecting seasonal changes in habitat use. The same bat should not be tagged twice in the course of one year unless there is a specific reason and it is covered by a project licence. Ringing bats is usually the way to determine which bats have been previously captured. Advice on ringing can be found in Mitchell-Jones and McLeish (2004) or Natural England's guidance on trapping and marking bats (NE, 2013).

For habitat use and nightly activity patterns, bats should be tracked for a minimum of three nights post-capture, and tracking should continue on more nights if the bat's movements do not become regular/consistent. A strong indication that sufficient data has been obtained is when cumulative plots of the study animal's home ranges reaches asymptote (for further information see Kenward, 2001).

From a survey planning perspective it is recommended that at least five tracking nights (post-capture) are planned for each bat to take account of bad weather or tag failures to ensure at least three nights' data can be obtained. Although useful in understanding the tagged bat's general activity patterns and to locate roosts, data from the capture night should not be used for later analysis of habitat use as the bat may be behaving differently due to the disturbance.

If bats are being marked with the objective of finding roosts, then it is advisable to continue to monitor the bat's roost movements for the lifetime of the transmitter, which can commonly be for up to two weeks. This level of monitoring will provide useful information on other roosts in the area, including night roosts.

9.3.8 Weather conditions

Radio tagging is usually associated with trapping bats from either field locations or at the roost. Please refer to Sections 2.6.1 and 9.2.8 for appropriate conditions. Tracking bats with radio transmitters generally does not suffer the environmental limitations of other survey methods as the survey is wholly reliant on the behaviour and activity pattern of the bats being tracked. There are numerous examples of radio-tracked bats flying in theoretically poor weather conditions for bats, especially when breeding or mating.

9.3.9 Next steps

Radio tagging and tracking is usually the last in a range of methods that might be used to determine the use of a proposed development site by bats. However, where roosts are discovered through radiotelemetry, then it may be necessary to carry out roost inspection surveys (see Chapters 5 and 6) or emergence/re-entry surveys (see Chapter 7).

Chapter 10

Data analysis and interpretation

10.1 Introduction

Data collected during bat surveys requires appropriate analysis, interpretation and presentation. The type of data collected depends on the surveys that were completed and what the aims and objectives of those surveys were. Where multiple surveys are proposed, it is good practice to analyse the data from the early surveys immediately to inform the later surveys, which may need to be adjusted according to the survey results. Analysing data at the end of a suite of surveys means that such opportunities would be missed.

Some examples of how to analyse, interpret and present bat data collected for proposed development projects are presented in the following sections.

10.2 Bat echolocation call analysis

10.2.1 General

The first stage of data processing is to complete sound analysis of bat calls. Russ (2012) provides detail about bats and sound equipment, call analysis and species identification and is a useful reference guide. A little information about software and species identification is provided below.

In the reporting of acoustic bat activity surveys, bat activity is often quantified in terms of numbers of bat passes but it can also be reported in terms of number of bat pulses (Sowler and Middleton, 2013). **It is important that the criterion for determining a bat pass is the same across all recordings that will be subject to comparison and that this criterion is reported.**

It is important to acknowledge that a bat pass or a bat pulse is an index of bat activity rather than a measure of number of individuals in a population. One hundred bat passes could be from 100 bats passing the detector or one bat passing 100 times. Reality is likely to fall somewhere between the two and this is where observational data can add context. There is little evidence that higher levels of bat echolocation activity actually reflect higher bat abundance (Hayes, 2000). Bat activity indices can be more accurately described as indices of the amount of use bats make of an area, and should be used to quantify bat activity, not abundance.

The benefit of recording bat activity is that there is an auditable record of work carried out. Bat echolocation data collected during bat surveys should be stored in case this auditable record requires later scrutiny.

10.2.2 Software

A number of sound analysis software options are available for both manual and automated sound analysis. Some software is brand-specific and can only handle recordings from specific bat detectors; other software is generic and can be used with a wide range of bat detectors. Choice of equipment and sound analysis software is likely to depend on the volume of data collected. Manual analysis may be appropriate for a small number of echolocation calls collected during an emergence survey. However, ecologists are increasingly collecting very large data sets (many thousands of bat calls) using automated/static detectors and automated analysis may be a more effective and efficient choice to handle the large volume of data and achieve consistency across a data set and between data sets.

The limitations of any sound analysis method used should be recognised and when using manual and/or automated methods, a proportion of the resulting data should always be verified for quality assurance purposes. A good approach with automated techniques is to at least verify all non-*Pipistrellus* calls manually and seek peer review where calls appear to be from rarer species, particularly if the site is outside their known range.

Regardless of the detecting equipment and software used, it is essential that an ecologist has the appropriate knowledge and experience to use it or results could be impaired.

10.2.3 Species identification

Bat call identification is difficult, even in the UK where there is a limited range of species. Some species, such as the greater and lesser horseshoe bats, can be identified with certainty from a spectrogram due to their unique call characteristics. Other species, for example a whiskered bat, can only be identified with a low degree of confidence to the species level but can be identified with a higher degree of confidence to its genus, *Myotis*.

The complexity involved in identifying bat calls is compounded by variability within the calls used by different species of bats. All species of bat vary the characteristics of their calls (e.g. frequency, call duration, inter-pulse interval) within a given range that is typical of the species. However, there is often a substantial degree of overlap for some or all characteristics between species. Calls are adapted dependent on behaviour (e.g. commuting, searching or approaching prey) and the surrounding habitat (e.g. in open or closed habitats or enclosed spaces) (see, for example, Holderied *et al.*, 2006; Murray *et al.*, 2001).

In addition to echolocation calls, bats also employ a wide range of social calls, which can be used to aid identification of bat species and to interpret their behaviour. More on interpreting social calls can be found in Middleton *et al.* (2014).

The quality of recorded calls will also depend on the location of the bat detector and the orientation of the bat to the microphone. If a detector is deployed on a hedge and a bat is flying over the hedge or behind the hedge the quality of the recorded call is likely to be lower than if the bat is echolocating directly at the microphone with no obstacles between the two to impede the passage of sound. Frequency has a big effect on how far away a call can be detected: lower-frequency calls can be detected from further away than higher-frequency calls. Most detectors will record bat calls at optimum quality (and at greater distances) if the call is received by the microphone in the same line as the long dimension of the microphone (on axis), although this is less important for omnidirectional microphones.

A proportion of bat species cannot be identified with certainty from their echolocation calls (sometimes due to the quality of the call but also because of the overlap in call characteristics between species) and it is important to consider and document how bats have been identified, either as single species or to genus (e.g. *Myotis*) or group (e.g. *Nyctalus/Eptesicus* or *Myotis/Plecotus*) and what level of confidence can be applied to identification. Automated sound analysis systems provide a level of confidence. Sometimes calls recorded are of insufficient quality to identify to any level and may be categorised as unknown bat calls.

When reporting results, it is always important to remember that different species vary in the likelihood of detection using bat detectors (Fenton, 1970) and it is therefore not relevant to compare numbers of bat passes/pulses from different species.

Because of the complexities outlined above, ecologists carrying out sound analysis should have a thorough understanding of how bat echolocation works and how call parameters can vary or the accuracy of the sound analysis could be impaired.

10.3 Analysis of bat activity survey data

Most types of bat surveys do not require statistical analysis. This section applies primarily to data collected during static/automated bat detector surveys (see Section 8.2). Statistical testing can be applied to other types of surveys (indeed, an example is given in Appendix 8), but it is only essential where large amounts of data are generated as it is otherwise difficult to extract meaning from the results. Analysis increases our understanding of the significance of differences in species composition and activity levels both spatially and temporally, which facilitates a more effective impact assessment. Statistics can be used to organise, summarise and describe the quantifiable data and can help to draw inferences in a transparent and authoritative way. The consequences of not undertaking formal statistical analyses are that some of the conclusions drawn from the data could be describing random 'noise' rather than something of statistical significance.

Data analysis is an iterative process by which data collected during field surveys and generated through the analysis of sound files recorded by bat detectors becomes knowledge and insight. The collation of data will involve cleaning the information for input errors, outliers, mistyping and highlighting missing values; for a protocol on how to achieve this, see Zuur *et al.* (2010).

Following data collation, a circular process of data transformation, visualisation and modelling takes place, as follows:

- **Transformation** is when data is manipulated and/or aggregated, creating new variables. One example of this is standardising bat activity observed per night through the season (with different night-time lengths) to activity per hour.
- **Visualisation** offers awareness of patterns within the data and uncovers the unexpected. However, it does not provide a scale to measure against, i.e. it does not clarify whether the differences in the data are random or significant.
- **Modelling** is where hypotheses are tested with statistical procedures (although some modelling techniques may not explicitly include hypotheses) to provide a scale to measure against, i.e. inference can be made about whether the differences in the data are random or significant.

Although data exploration is a key part of any analysis, it is recommended that it is clearly separated from hypothesis testing. It is good practice to decide what statistical tests to apply during the survey design (or after a 'pilot study' or initial survey), i.e. they should be decided **before the surveys** based on the ecologist's understanding of the questions being asked and their biological understanding of the system. Data analysis should be viewed as an aid to the decision-making process that has followed through from the objective of the survey and survey design (Underwood, 1997) (see Section 2.2.6).

Table 10.1 names some statistical tests that can be applied to bat survey data. The tests listed are robust in that the observed data can be used as it comes, and no assumption is made about the distribution of the data; all the tests are what are termed non-parametric. There are some minimum requirements, which are detailed in Table 10.1.

Before using any of the statistical tests it is recommended that you refer to a reference that gives a background on how to apply the test and its limitations. There are many other statistical procedures that can be applied to ecological data; Dytham (2011) provides an introductory text and Legendre and Legendre (2012) provide more detail.

The statistical tests listed look for differences and/or relationships and are helpful in interpreting bat survey data for reporting; the tests are useful in separating the signal from the noise. There are simple tests to look at the differences for individual species and multivariate tests that allow for the comparison of communities (e.g. the assemblage of bats) from species 'abundance' data. These tests also allow for ecological data to be explored against environmental factors. All the statistical tests add weight to professional opinion.

Table 10.1 Statistical tests that can be applied to bat survey data.

Example application of data analysis	Statistical test
Compare two samples of bat activity (expressed as bat passes per night for an individual species or groups of species). For example, bat activity observed at one location over several nights in June (sample 1) compared with activity observed at the same location over several nights in July (sample 2). These data are unpaired. OR bat activity observed at two locations, such as a hedge (sample 1) and an open field (sample 2) over several nights. These data are paired.	Non-parametric tests that look at the differences for individual species (see Dytham, 2011). Mann–Whitney U test: (for unpaired data) [the number of nights in each sample can be different; see Fowler *et al.*, 1998]. Wilcoxon's signed rank test: (for paired data) [the number of matched pairs whose difference is not zero should be six or more; see Fowler *et al.*, 1998].
Compare *three or more samples* of bat activity (expressed as bat passes per night for an individual species or groups of species). For example, bat activity observed at one location over several nights in each of June (sample 1), July (sample 2), August (sample 3) and September (sample 4). OR bat activity observed at three or more locations (each one is a sample) over several nights in only one month.	Non-parametric test that looks at the differences for individual species (see Dytham, 2011). Kruskal–Wallis test: [if there are only three samples then there must be at least five nights in each sample; see Fowler *et al.*, 1998]. As above.
Test *three or more samples* of bat activity for whether the order of the samples is meaningful; is there a decreasing or increasing trend (data expressed as bat passes per night for an individual species or groups of species)? For example, bat activity observed at one location over several years 2010, 2011, 2012, 2013, 2014.	Non-parametric test that looks at trends in individual species (see Field *et al.*, 2012). Jonckheere–Terpstra test: [this is similar to the Kruskal–Wallis test but looks for information about whether the order of the samples or groups is meaningful – so it can test for a decreasing or increasing trend].

The skill and resources required for managing data and undertaking data analysis should not be underestimated. Bat survey projects can be undertaken over many years and it is not uncommon for the project team to change during this time; it is therefore good practice to manage information so others can understand and have access to what has been done. This requires the management and analysis of data to be transparent and reproducible by others. There are software tools that make the process of data management, analysis and reporting reproducible; many of the software tools to undertake this are open source and available for all to use (see Box 6).

Box 6 Tools for data management, analysis and reproducible reporting.

Data management: Excel® (https://products.office.com/en-gb/excel) and its Open Office (https://www.openoffice.org/) equivalent is a useful data management tool.

Data analysis: a powerful open source statistical software environment is available with R (https://cran.r-project.org/). Commercially available data analysis software includes SPSS (http://www-01.ibm.com/software/uk/analytics/spss/); Minitab (http://www.minitab.com/); SAS (http://www.sas.com/); STATA (http://www.stata.com/) and software aimed at biologists and ecologists Primer-E (http://www.primer-e.com/).

Reproducible reporting: the open source Integrated Development Environment (IDE) RStudio™ (https://www.rstudio.com/) with its implementation of R and RMarkdown (http://rmarkdown.rstudio.com/) enables ecologists to gather data (from Excel or Open Office) and visualise and run statistical analyses. Through RStudio™ you can connect the R-based analysis dynamically and reproducibly to presentations and reports; created in mark-up languages such as Markdown and LaTeX (http://www.latex-project.org/). Directly linking your data, your analyses, and your results, a process called literate programming (Knuth, 1984), makes tracing your steps much easier.

Appendix 7 gives an introduction to data analysis, describing some simple transformation, visualisation and modelling of data and some worked examples are provided in Appendix 8. The modelling/analysis described is mostly non-parametric, which makes fewer assumptions about the data, is simple to apply and is suited to analysing the large and small samples that are frequently found with ecological data. All of the examples provided in Appendix 7 and Appendix 8 have been created in R.

10.4 Analysis of bat radiotelemetry survey data

This section applies primarily to data collected during radiotelemetry surveys (see Section 9.3). For a detailed account of radiotelemetry and analyses of radiotelemetry data please see Kenward (2001). Some of the common analysis techniques associated with radiotelemetry and bats are given below.

Establishing home ranges is particularly useful in understanding the extent of use of a proposed development site in relation to the surrounding landscape. This is usually an area-based calculation determined after tracking the bat for a period of time that establishes a regular pattern of activity. From home range calculations, it may be possible to determine what proportion of the home range of the bat or colony the proposed development site is likely to comprise.

Bats often move through large areas to spend time foraging or socialising in smaller 'core' areas. It is often important to quantify these core areas, as overall home ranges do not necessarily determine the 'important' areas/habitats that are used by the bat.

There are a number of methods for estimating the home ranges and core areas of bats. The common methods are minimum convex polygons (MCP), cluster analysis and kernel contours. Kenward (2001) provides detail on all the main methods. However, it should be noted that the selection of the home range estimation tool should be appropriate for the behaviour of the bat. Some bats (e.g. Bechstein's bat) may make small movements from roost to foraging areas and the selection of kernel contours might be appropriate, whereas for fast-moving bats that use discrete foraging sites scattered across the landscape, the use of cluster analysis and MCPs would be more appropriate.

For studies that are seeking to determine habitat preferences of the bats affected by a development proposal, it is important to use statistical techniques to quantify and establish such preferences. A common method of analysis of habitat selection is to compare the proportion of habitats used by the bats the majority of the time (i.e. core areas) to the habitats that were available to the bat within its home range (MCP). Habitat selection of areas used versus available can be determined through the use of statistical tests such as the compositional analysis methods developed by Aebischer *et al.* (1993). To be reliable, these methods require an understanding of where each bat was located for a significant proportion of each night tracked, and is more difficult for fast-moving bats.

It is also important that appropriate habitat data is collected covering the areas available to the bat (e.g. the MCP) (see Section 9.3.7 on survey effort).

Chapter 11

Writing bat reports

11.1 Introduction

It is essential that bat survey reports are accurate, clear, concise and, most importantly, serve the purpose for which they were intended. A survey report for the purposes of these guidelines is reporting on what is there and may be making recommendations for action. A monitoring report is reporting on what action has been taken, whether it has been implemented correctly and whether it has been effective. Reporting on monitoring is not covered by these guidelines.

This chapter covers the essentials of good bat survey report writing and provides a standardised template for a bat survey report. Information can also be found in *Guidelines for Ecological Report Writing* (CIEEM, 2015).

Put simply, a bat survey submitted in support of a development proposal should show:

- what is there and its value and significance;
- how it will be impacted by the development;
- how these impacts can be mitigated;
- how the development will result in no net loss (and where possible a net gain – particularly for planning purposes) to their population.

In general, professional reports should:

- be accessible to the intended audience;
- use clear and simple sentence structures;
- be proofread for grammar, spelling and punctuation;
- list both scientific and common names of species;
- cite appropriate references to back up assertions;
- use a standard, consistent format for references;
- leave no room for misinterpretation; and
- propose clear, definitively stated actions resulting from the findings of the report.

11.2 Standard template for bat survey reports

Box 7 provides guidelines on the content of individual sections of a bat survey report produced in relation to planning and development. It may be possible to streamline the process of report writing by producing reports that are fit for multiple purposes. Not all sections are relevant in all situations; professional judgement should be used in determining the final format.

Box 7 Sections and content relevant to bat survey reports for planning and development.

Title page

- Concise title explaining the type of survey, the subject of the survey and the location, e.g. 'Preliminary Roost Assessment of Barn at Brook Farm'.
- The date and version number of the report.
- The client's name and/or organisation.
- The author's name and/or organisation.
- Other relevant information such as 'draft' or 'confidential'.

Executive summary

- A non-technical, concise summary of the whole report including the purpose of the report, the site context, survey methods, survey results, limitations and methods to overcome limitations, further survey recommendations, impact assessment, methods to avoid, mitigate or compensate, enhancement measures, post-construction monitoring measures and conclusions as appropriate. This should be self-contained and may not be needed if the report itself is very short.

Contents page

- List of sections including numbers, titles and page numbers.
- List of all figures, tables, graphs and photographs including numbers, titles and page numbers.

Introduction

- Purpose/context of the report: written by whom, for whom and why.
- Proposed development activities, including future use of site. If not known, this should be stated.
- Site context – size, brief description, brief description of habitat, locational information (description, grid reference, postcode), map showing site boundary, aerial photographs, photographs.

- Brief description of surveys carried out including aims and objectives.
- Reference to other reports or information available prior to the surveys being carried out, e.g. preliminary ecological appraisal or reports from other ecologists.

Methods

- Desk study: a list of organisations and sources from which designated sites and bat records were requested and obtained, how the search area was specified; the date that the search was made; reasons for not carrying out a data search if relevant.
- For each type of bat survey carried out and for each separate survey occasion (where relevant):
  - bat survey types used;
  - equipment/software used;
  - description of method (including how bat pass was defined and parameters used for echolocation analysis);
  - justification for choice of method and equipment (linking to aims and objectives) including any deviation from good practice (reference these guidelines) and rationale;
  - how the design of the survey was informed by previous surveys (or by the desk study);
  - number of ecologists;
  - ecologist names;
  - relevant ecologist training, experience, licences and licence numbers;
  - area surveyed with justification for choice of survey area and maps/aerial photographs for reference;
  - all ecologist and equipment locations (e.g. emergence/dawn ecologist locations, transect routes, static survey locations using automated detectors, location of mist nets and harp traps) for each separate survey, with justification for choice of locations and maps/aerial photographs for reference;
  - survey dates;
  - survey start and end times;
  - sunset/sunrise times;
  - limitations of survey methods (e.g. weather, access, timing, health and safety considerations) or equipment.

Results

- Preliminary ecological appraisal – desk study: a list of sites designated for their bat interest plus descriptions and a summary of bat species and roosts in the area, with a map if available/relevant/possible (the amount of detail provided will depend on the terms and conditions of the data provider).
- Preliminary ecological appraisal – fieldwork: a Phase 1 map with target notes describing and assessing suitability of features for roosting, foraging and commuting bats; a set of photographs of the site.
- Preliminary roost assessment of structures and winter hibernation surveys:
  - descriptions of structures surveyed (including reference number, location, type of building/structure, dimensions, age, construction materials, current use);
  - descriptions of potential and actual access points and roosting places (including height above ground level and aspect);
  - descriptions of evidence of bats found;
  - results of DNA analysis undertaken;
  - description of areas not surveyed and reasons why;
  - all of the above marked onto a plan of the site;
  - a set of cross-referenced photographs highlighting key features.
- Preliminary ground level roost assessment of trees:
  - descriptions of trees surveyed (including reference number, species, diameter at breast height);
  - descriptions of potential and actual roost features (including height above ground level and aspect);
  - description of evidence of bats found;
  - trees not surveyed and reasons why;
  - all of the above marked onto a plan of the site;
  - a set of cross-referenced photographs.
- PRF inspection survey – trees
  - description of potential and actual roost features surveyed (including dimensions, level of protection from elements);
  - description of evidence of bats found;
  - features not surveyed and reasons why;
  - all of the above marked onto a plan of the site;
  - a set of cross-referenced photographs.
- Presence/absence and roost characterisation surveys:
  - descriptions of emerging/returning bats (including time, species, number, exit/entry point, behaviour observed);
  - descriptions of other notable bat behaviour (including internal flight, observations of major commuting routes locally);
  - all of the above marked onto a plan of the site.
- Bat activity surveys:
  - tables of bats recorded/observed (including time, species, number of passes, behaviour observed) where low numbers or this information summarised where higher numbers recorded;
  - the above information summarised on an annotated plan or aerial photograph of the site.

- ALBST (minimum data required):
  - tables of bats captured in relation to trap locations (including time/date, species, age class, breeding status and any other data collected);
  - tables of radio-tracked bat summary data to include tracking dates, number of nights tracked, number of fixes obtained for each bat, home range size and maximum distance from roost.

Evaluation

- Data visualisation, analysis and interpretation of the results. This section is particularly important because it links the results of the surveys with the impact assessment and subsequent recommendations. There should be enough information to make this link explicit.
- Limitations of survey (with respect to weather, survey methods, timing, equipment, detectability of different species, etc.) and impacts on survey results.
- Relevant European and UK legislation, relevant national and local planning policy, national and local bat species biodiversity action plans. Place the findings of the survey into a legal and policy context.

Impact assessment

- Assessment of the impacts of the proposed development pre- and during construction and during operation and decommissioning (where relevant) on designated sites, roosts and commuting and foraging areas used by bats.

Required actions

- Further surveys – exact requirements described.
- Justification on the necessity or otherwise for an EPS licence to be obtained.
- Avoidance, mitigation, compensation and enhancement measures. All measures should be quantified, definitively stated, marked onto diagrams and drawn up in consultation with the client. Language such as 'should' and 'could' must not be used to describe a required measure. Instead, use 'will', as long as this has been agreed with a client (this may not be possible in early iterations of a series of reports). This enables planners to impose clear, enforceable conditions relating to this section of the report.
- Post-construction monitoring. See comments above on enforceability and use of language.

References

Glossary or definition of terms.

Appendices

- Should include supplementary or supporting material that would interrupt the flow of the main report. May include maps, aerial photographs, GIS files (which can be useful for large and complex schemes), figures, photographs and background/raw data.

11.3 Use of illustrative material

The importance of illustrative material in reports should not be underestimated. A report should convey the required information in the most concise and easy to understand format – an annotated map, aerial photograph, diagram, graph, figure or photograph can replace many words. Examples of how to visualise data are provided in Appendix 7 and Appendix 8.

11.4 Other considerations

11.4.1 Peer reviewing

Professional reports should not be sent out without a peer review, generally by a more senior or experienced colleague. This identifies any errors with grammar, spelling and punctuation but also ensures that the content is appropriate for the audience and the recommendations are clear and justified. Many consultancies have a good practice system for signing off reports where the author and the reviewer are identified and signatures are required for final approval and submission.

11.4.2 Submission of bat records

It is good practice for ecologists to state in their terms and conditions that records from surveys will be submitted to record-holding organisations. Bat records can then be submitted to Local Records Centres, LBGs or the NBN.[46]

In Northern Ireland the ecologist has a choice of who they submit their data to:

1. Northern Ireland Bat Group; **OR**
2. Centre for Environmental Data and Recording (CEDaR), which is Northern Ireland's Local Record Centre; **OR**
3. National Biodiversity Data Centre (NBDC) in the Republic of Ireland that hosts the 'Atlas of Irish Mammals' for both Irish jurisdictions and shares all relevant records with CEDaR (above).

This practice should be encouraged, for the benefit of all stakeholders, and only waived in exceptional circumstances where there is genuine justification.

[46] http://www.searchnbn.net

References

Adams, A.M., Jantzen, M.K., Hamilton, R.M. and Fenton, M.B. (2012) Do you hear what I hear? Implications of detector selection for acoustic monitoring of bats. *Methods in Ecology and Evolution* 3: 992–998.

Aebischer, N.J., Robertson, P.A. and Kenward, R.E. (1993) Compositional analysis of habitat use from animal radiotracking data. *Ecology* 74: 1313–1325.

Altringham, J. D. (2003) *British Bats*. Collins New Naturalist Library, Volume 93. Harper Collins, London.

Andrews, H. (2013) Bat Tree Habitat Key. Available from: http://www.aecol.co.uk/Pages/48/Research—Development—Bat-Tree-Habibat-Key.html (accessed 24 November 2015).

Andrews, H. and Gardener, M. (2015) Surveying trees for bat roosts: Encounter probability v. survey effort. *In Practice* 88: 33–38.

Angell, R.L., Butlin, R.K. and Altringham, J.D. (2013) Sexual segregation and flexible mating patterns in temperate bats. *PLoS ONE* 8(1): e54194. doi:10.1371/journal.pone.0054194.

Barlow, K.E. and Jones, G. (1999) Roosts, echolocation calls and wing morphology of two phonic types of *Pipistrellus pipistrellus*. *Zeitschrift fur Säugertierkunde* 64: 257–268.

Bat Conservation Trust (2012) *Professional Training Standards for Ecological Consultants*. BCT, London.

Bat Conservation Trust (2013) *Bechstein's Bat: An introduction for woodland managers*. BCT, London. Available from: http://www.bats.org.uk/data/files/Bechsteins_bat_woodland_guide_2nd_Ed_2013.pdf (accessed 24 November 2015).

Bat Conservation Trust/BMT Cordah Limited (2005) *A Review and Synthesis of Published Information and Practical Experience on Bat Conservation within a Fragmented Landscape*. An occasional report by The Three Welsh National Parks, Pembrokeshire CC and the Countryside Council for Wales, Cardiff.

Berge, L. (2007) Resource partitioning between the cryptic species Brandt's bat (*Myotis brandtii*) and the whiskered bat (*Myotis mystacinus*). PhD Thesis. University of Bristol, Bristol.

Berthinussen, A. and Altringham, J. (2015) *Development of a Cost-Effective Method for Monitoring the Effectiveness of Mitigation for Bats Crossing Linear Transport Infrastructure – WC1060*. University of Leeds, Leeds. Available from: http://sciencesearch.defra.gov.uk/Default.aspx?Menu=Menu&Module=More&Location=None&Completed=0&ProjectID=18518 (accessed 3 December 2012).

Billington, G.E. (2000) *Preliminary Report on Further Research on the Barbastelle at Holnicote Estate*. English Nature, Peterborough.

Billington, G.E. (2003) *Radio-tracking study of Barbastelle bats in Pengelli Forest National Nature Reserve*. CCW Contract Science Report No. 590 [confidential]. Countryside Commission Wales, Bangor.

Billington, G.E. and Norman, G.M. (1997) *The Conservation of Bats in Bridges Project: A Report on the 1996 Survey and Conservation of Bat Roosts in Bridges in Cumbria*. English Nature, Peterborough. Available from: http://publications.naturalengland.org.uk/publication/5454603824922624 (accessed 24 November 2015).

Bilston, H. (2014) *Maximising Occupation of Bat Boxes in an Ancient Woodland in Buckinghamshire: A Summary of Recent Research*. BSG Ecology, Oxford.

Bontadina, F., Schofield, H. and Naef-Daenzer, B. (2002) Radio-tracking reveals that lesser horseshoe bats (*Rhinolophus hipposideros*) forage in woodland. *Journal of Zoology, London* 258: 281–290.

Boonman, M. (2000) Roost selection by noctules (*Nyctalus noctula*) and Daubenton's bats (*Myotis daubentonii*). *Journal of Zoology, London* 251: 385–389.

Briggs, P. (1995) Bats in barns. *Transactions of the Hertfordshire Natural History Society* 32: 237–244.

Briggs, P. (2002) *A study of bats in barn conversions in Hertfordshire in 2000*. Hertfordshire Biological Records Centre, Hertford.

BSG Ecology (2013a) *Dungeness, Kent: Bat Migration Pilot Study*. BSG Ecology, Worton, Oxfordshire. Available from: http://www.bsg-ecology.com/wp-content/uploads/2015/01/Dungeness-Bat-Study-2012.pdf (accessed 24 November 2015).

BSG Ecology (2013b) *Kent Bat Migration Research: Baseline Report*. BSG Ecology, Worton, Oxfordshire. Available from: http://www.bsg-ecology.com/wp-content/uploads/2015/01/Kent-Bat-Migration-Research-Baseline-Report__12122013.pdf (accessed 4 December 2015).

BSG Ecology (2013c) *Spurn Bat Migration Study*. BSG Ecology, Worton, Oxfordshire. Available from: http://www.bsg-ecology.com/wp-content/uploads/2015/01/Spurn-Bat-Report-2013.pdf (accessed 6 December 2012).

BSG Ecology (2013d) *Portland Bat Migration Study*. BSG Ecology, Worton, Oxfordshire. Available from: http://www.bsg-ecology.com/wp-content/uploads/2015/01/Portland-Bat-Summary-2013-Final-NV.pdf (accessed 6 December 2015).

BSG Ecology (2014a) *North Sea Ferries Bat Migration Research Report*. BSG Ecology, Worton, Oxfordshire. Available from: http://www.bsg-ecology.com/wp-content/uploads/2015/01/North-Sea-Ferry-Bat-Migration-Research-Report-20141.pdf (accessed 4 December 2015).

BSG Ecology (2014b) *Bat Migration Research Report: Stable Isotope Analysis of Nathusius' Pipistrelle Fur Samples*. BSG Ecology, Worton, Oxfordshire. Available from: http://www.bsg-ecology.com/wp-content/uploads/2015/01/SIA_APPR_2711141.pdf (accessed 4 December 2015).

BSI (2013) *BS42020:2013 Biodiversity. Code of practice for planning and development*. British Standards Institution, London.

BSI (2015) *BS8596:2015 Surveying for bats in trees and woodland*. British Standards Institution, London, UK.

Buckley, D.J., Lundy, M.G., Boston, E.S.M., Scott, D.D., Gager, Y., Prodohl, P., Marnelle, F., Montgomery, W.I. and Teeling, E.C. (2013) The spatial ecology of the whiskered bat (*Myotis mystacinus*) at the Western extreme of its range provides evidence of regional adaptation. *Mammalian Biology* 78: 198–204.

Catto, C.M.C., Hutson, A.M., Racey, P.A. and Stephenson, P.J. (1996) Foraging behaviour and habitat use of the serotine bat (*Eptesicus serotinus*) in southern England. *Journal of Zoology, London* 238: 623–633.

CIEEM (2013a) *Competencies for Species Surveys: Bats*. Chartered Institute of Ecology and Environmental Management, Winchester. Available from: http://www.cieem.net/data/files/Resource_Library/Technical_Guidance_Series/CSS/CSS_-_BATS_April_2013.pdf (accessed 4 December 2015).

CIEEM (2013b) *Code of Professional Conduct*. Chartered Institute of Ecology and Environmental Management, Winchester. Available from: http://www.cieem.net/data/files/Resource_Library/Professional_Conduct/Code_of_Professional_Conduct_June_2013.pdf (accessed 4 December 2015).

CIEEM (2013c) *Good Working Practices*. Chartered Institute of Ecology and Environmental Management, Winchester. Available from: http://www.cieem.net/data/files/Resource_Library/Professional_Guidance_Series/Good_Working_Practices_FINAL.pdf (accessed 6 December 2015).

CIEEM (2013d) *Guidelines for Preliminary Ecological Appraisal*. Chartered Institute of Ecology and Environmental Management, Winchester. Available from: http://www.cieem.net/data/files/Resource_Library/Technical_Guidance_Series/GPEA/GPEA_April_2013.pdf (accessed 4 December 2015).

CIEEM (2015) *Guidelines for Ecological Report Writing*. Chartered Institute of Ecology and Environmental Management, Winchester. Available from: http://www.cieem.net/data/files/Resource_Library/Technical_Guidance_Series/Guidelines_for_Ecological_Report_Writing/Guidelines_for_Ecological_Report_Writing_and_Appendices_May2015.pdf (accessed 6 December 2015).

CIEEM (2016) *Guidelines for Ecological Impact Assessment in the UK and Ireland: Terrestrial, Freshwater and Coastal* (2nd edn). Chartered Institute of Ecology and Environmental Management, Winchester.

Clarke, K.R. (1993) Non-parametric multivariate analysis of changes in community structure. *Australian Journal of Ecology* 18: 117–143.

Corbet, G.B. and Harris, S. (1991) *The Handbook of British Mammals* (3rd edn). Blackwell Scientific Publications, Oxford.

Davidson-Watts, I. (2007) Roost selection, foraging behaviour and habitat use by two cryptic species of pipistrelle bat. PhD thesis. Open University/University of Bristol.

Davidson-Watts, I. (2008) *The Isle of Wight Woodland Bat Project: Final Report*. People's Trust for Endangered Species, London.

Davidson-Watts, I. (2013) *Briddlesford Copse Bechstein's and Barbastelle Bat Project 2013-2016 YEAR 1*. Davidson-Watts Ecology Limited. Unpublished report, available from the author at: http://www.dwecology.co.uk

Davidson-Watts, I. (2014a) *Barbastelle bat surveys and tracking at Nocton wood, Lincolnshire*. Davidson-Watts Ecology Limited. Unpublished report, available from the author at: http://www.dwecology.co.uk

Davidson-Watts, I. (2014b) *Advanced bat surveys of land east of Horndean, Hampshire, draft report*. Davidson-Watts Ecology Limited. Unpublished report, available from the author at: http://www.dwecology.co.uk

Davidson-Watts, I. and Jones, G. (2006) Differences in foraging behaviour between *Pipistrellus pipistrellus* (Schreber, 1774) and *Pipistrellus pygmaeus* (Leach, 1825). *Journal of Zoology, London* 268, 55–62.

Davidson-Watts, I., Walls, S. and Jones, G. (2006) Differential habitat selection by *Pipistrellus pipistrellus* and *Pipistrellus pygmaeus* identifies distinct conservation needs for cryptic species of echolocating bats. *Biological Conservation* 133: 118–127.

DCLG (1999) *Circular 02/99: Environmental impact assessment*. Department for Local Communities and Government, London. Available from: https://www.gov.uk/government/uploads/system/uploads/attachment_data/file/7676/155958.pdf (accessed 5 December 2015).

DCLG (2005) *Government Circular: Biodiversity and geological conservation – Statutory obligations and their impact within the planning system*. Department for Communities and Local Government, London. Available from: https://www.gov.uk/government/uploads/system/uploads/attachment_data/file/7692/147570.pdf (accessed 5 December 2015).

DCLG (2012) *National Planning Policy Framework*. Department for Communities and Local Government, London. Available from: https://www.gov.uk/government/uploads/system/uploads/attachment_data/file/6077/2116950.pdf (accessed 5 December 2015).

Dietz, C., Helversen, O. and Dietmar, N. (2011) *Bats of Britain, Europe & Northwest Africa*. A&C Black, London.

DOENI Planning and Environmental Policy Group (2009) *Planning Policy Statement 18: 'Renewable Energy'*. Department of the Environment Northern Ireland, Belfast. Available from: http://www.planningni.gov.uk/index/policy/planning_statements_and_supplementary_planning_guidance/planning_policy_statement_18__renewable_energy.pdf (accessed 5 December 2015).

DOENI Planning Policy Division (2013) *Planning Policy Statement 2: 'Natural Heritage'*. Department of the Environment Northern Ireland, Belfast. Available from: http://www.planningni.gov.uk/index/policy/planning_statements/definitive_final_july_2013_pps_2_-_natural_heritage-3.pdf (accessed 5 December 2015).

Duvergé, P.L. and Jones, G. (1994) Greater horseshoe bats – activity, foraging behaviour and habitat use. *British Wildlife* 6: 69–77.

Duvergé, P.L. and Jones, G. (2003) Use of farmland habitats by greater horseshoe bats. In: *Farming and Mammals. Conservation and Conflict* (Tattersall, F. and Manley, W., eds). The Linnaean Society, London, pp. 64–81.

Dytham, C. (2011) *Choosing and Using Statistics: A Biologist's Guide* (3rd edn). Wiley-Blackwell, Chichester.

English Nature (2003) *Managing landscapes for the greater horseshoe bat*. English Nature, Peterborough. Available from: http://adlib.everysite.co.uk/adlib/defra/content.aspx?doc=123742&id=124125 (accessed 5 December 2015).

Entwistle, A.C., Racey, P.A. and Speakman, J.R. (1996) Habitat exploitation by a gleaning bat, *Plecotus auritus*. *Philosophical Transactions of the Royal Society of London B* 351: 921–931.

Entwistle, A.C., Racey, P.A. and Speakman, J.R. (1997) Roost selection by the brown long-eared bat *Plecotus auritus*. *Journal of Applied Ecology* 34: 399–408.

Faure, P.A., Fullard, J.H. and Barclay, R.M.R. (1990) The response of tympanate moths to the echolocation calls of a substrate gleaning bat, *Myotis evotis*. *Journal of Comparative Physiology A* 166: 843–849.

FC England, FC Wales, BCT, CCW and English Nature (2005) *Woodland management for bats*. Forestry Commission, Bristol. Available from: http://www.forestry.gov.uk/pdf/woodland-management-for-bats.pdf/$FILE/woodland-management-for-bats.pdf (accessed 5 December 2015).

Fenton, M.B. (1970) A technique for monitoring bat activity with results obtained from different environments in southern Ontario. *Canadian Journal of Zoology* 48: 847–851.

Field, A., Miles, J. and Field, Z. (2012) *Discovering Statistics Using R*. Sage Publications, London.

Fitzsimons, P., Hill, D. and Greenaway, F. (2002) *Patterns of habitat use by female Bechstein's bats (Myotis bechsteinii) from a maternity colony in a British woodland*. School of Biological Sciences, University of Sussex, Sussex.

Flanders, J. and Jones, G. (2009) Roost use, ranging behaviour and diet of greater horseshoe bats (*Rhinolophus ferrumequinum*) using a transitional roost. *Journal of Mammalogy*, 90: 888–896.

Fowler, J., Cohen, L. and Jarvis, P. (1998) *Practical Statistics for Field Biology* (2nd edn). Wiley, Chichester.

Furmankiewicz, J. and Altringham, J.D. (2007) Genetic structure in a swarming brown long-eared bat (*Plecotus auritus*) population: evidence for mating at swarming sites. *Conservation Genetics* 8: 913–923.

Glover, A.M. and Altringham, J.D. (2008) Cave selection and use by swarming bat species. *Biological Conservation* 141: 1493–1504.

Goerlitz, H.R., Hofstede, H.M., Zeale, M.R.K., Jones, G. and Holderied, M.W. (2010) An aerial-hawking bat uses stealth echolocation to counter moth hearing. *Current Biology*, 20: 1568–1572.

Greenaway, F. (2001) The barbastelle in Britain. *British Wildlife* 12: 327–334.

Greenaway, F. (2008) *Barbastelle Bats in the Sussex West Weald 1997–2008*. A report to Sussex Wildlife Trust and the West Weald Landscape Project. Available from: http://www.westweald.org.uk/pdf/Barbastelle%20Bats%20in%20the%20Sussex%20West%20Weald%201997-2008.pdf (accessed 5 December 2015).

Greenaway, F. and Hill, D. (2004) *Woodland management advice for Bechstein's bat and barbastelle bat*. English Nature Research Report No. 658, English Nature, Peterborough.

Grubb, E.N. (2012) The effect of landscape scale weather conditions on swarming activity of Natterer's bat (*Myotis nattereri)*. MSc thesis, University of Leeds.

Gunnell, K., Murphy, B. and Williams, C. (2013) *Designing for Biodiversity: A technical guide for new and existing buildings* (2nd edn). RIBA Publishing, London.

Hargreaves, D. (2012) *Nathusius pipistrelle project 2011 summary report: Blagdon Lake, North Somerset*. A report for Yatton and Congresbury Wildlife Action Group. Available from: http://yacwag.org.uk/wp-content/uploads/2012/03/Nathusius-pipistrelle-tracking-reportV3Apr12.pdf (accessed 5 December 2015).

Hargreaves, D. (2014) *Nathusius Pipistrelles and BCT Survey*. The Mammal Society. Available from: http://www.mammal.org.uk/60for60_weekly_blogs/nathusius-pipistrelles (accessed 5 December 2015).

Harris, S. and Yalden, D.W. (2008) *Mammals of the British Isles: Handbook* (4th edn). The Mammal Society, Southampton.

Hayes, J.P. (2000) Assumptions and practical considerations in the design and interpretation of ecological monitoring studies. *Acta Chiropterologica* 2: 225–236.

Hill, D.A. and Greenaway, F. (2005) Effectiveness of an acoustic lure for surveying bats in British woodlands. *Mammal Review* 35: 116–122.

Holderied, M.W., Jones, G. and von Helversen, O. (2006) Flight and echolocation behaviour of whiskered bats commuting along a hedgerow: range-dependent sonar signal design, Doppler tolerance and evidence for 'acoustic focussing'. *Journal of Experimental Biology* 209: 1816–1826.

Hundt, L. (2012) *Bat Surveys: Good Practice Guidelines* (2nd edition). Bat Conservation Trust, London.

Hutson, A.M. (1984) *Keds, flat-flies and bat-flies. Diptera, Hippoboscidae and Nycteribiidae*. Handbooks for the Identification of British Insects, Vol. 10, Part 7. Royal Entomological Society, London. Available from: http://www.royensoc.co.uk/sites/default/files/Vol10_Part07_Hutson.pdf (accessed 5 December 2015).

JNCC (2007) *2nd UK Report on Implementation of the Habitats Directive*. Joint Nature Conservation Committee, Peterborough. Available from: http://jncc.defra.gov.uk/page-4060 (accessed 18 December 2015).

JNCC (2010) *Handbook for Phase 1 Habitat Survey – a Technique for Environmental Audit*. Joint Nature Conservation Committee, Peterborough. Available from: http://jncc.defra.gov.uk/page-2468#download (accessed 5 December 2015).

JNCC (2013) 3rd UK Habitats Directive Reporting 2013. Joint Nature Conservation Committee, Peterborough. Available from: http://jncc.defra.gov.uk/page-6387 (accessed 18 December 2015).

Jones, G. and Raynor, J.M.V. (1988) Flight performance, foraging tactics and echolocation in free-living Daubenton's bats *Myotis daubentonii* (Chiroptera, Vespertilionidae). *Journal of Zoology, London* 215: 113–132.

Kahle, D. and Wickham, H. (2013) ggmap: Spatial Visualization with ggplot2. *The R Journal* 5(1): 144–61. Available from: https://journal.r-project.org/archive/2013-1/kahle-wickham.pdf (accessed 5 December 2015).

Kenward, R.E. (2001) *A Manual for Wildlife Radio Tagging*. Academic Press, London.

Kerth, G., Weissman, K. and Konig, B. (2001) Day roost selection in female Bechstein's bats (*Myotis bechsteinii*): a field experiment to determine the influence of roost temperature. *Oecologia* 126: 1–9.

Kerth, G., Kiefer, A., Trappmann, C. and Weishaar, M. (2003) High gene diversity at swarming sites suggest hot spots for gene flow in the endangered Bechstein's bat. *Conservation Genetics* 4: 491–499.

Knuth, D. E. (1984) Literate programming. *The Computer Journal* 27: 97–111.

Korsten, E., Schillemans, M., Limpens, H. and Jansen, E. (2015) On the trail of the hibernating common pipistrelle (*Pipistrellus pipistrellus*): Strategies for locating mass hibernation sites in the urban environment. National Bat Conference 2015 abstracts, Bat Conservation Trust, London. Available from: http://www.bats.org.uk/data/files/National_Conference_Abstrcts_2015.pdf (accessed 18 December 2015).

Kronwitter, F. (1988) Population structure, habitat use and activity patterns of the noctule bat, *Nyctalus noctula Schreb.*, 1774 (Chiroptera: Vespertilionidae) revealed by radio-tracking. *Myotis* 26: 23–83.

Kunz, T. H. and Parsons, S. (eds) (2009) Ecological and behavioral methods for the study of bats. *Journal of Mammalogy* 92: 475–478.

Legendre, P. and Legendre, L. (2012) *Numerical Ecology* (3rd edn). Developments in Environmental Modelling, Volume 24. Elsevier, Amsterdam.

Lundberg, K. and Gerell, R. (1986) Territorial advertisement and mate attraction in the bat *Pipistrellus pipistrellus*. *Ethology* 71: 115–124.

Mackie, I.J. and Racey, P.A. (2007) Habitat use varies with reproductive state in noctule bats (*Nyctalus noctula*): Implications for conservation. *Biological Conservation* 140: 70–77.

Mathews, F., Roche, N., Aughney, T., Jones, N., Day, J., Baker, J. and Langton, S. (2015) Barriers and benefits: implications of artificial night-lighting for the distribution of common bats in Britain and Ireland. *Philosophical Transactions of the Royal Society of London B* 370(1667), doi: 10.1098/rstb.2014.0124.

Middleton, N., Froud, A. and French, K. (2014) *The Social Calls of the Bats of Britain and Ireland*. Pelagic Publishing, Exeter.

Miller, H. (2012) *Bechstein's Bat Survey: Final Report*. The Bat Conservation Trust, London. Available from: http://www.bats.org.uk/data/files/publications/Bechsteins_bat_survey_final_report.pdf (accessed 5 December 2015).

Mitchell-Jones, A.J. (2004) *Bat Mitigation Guidelines*. English Nature, Peterborough. Available from: http://www.warksbats.co.uk/pdf/Batmitigationguide.pdf (accessed 5 December 2015).

Mitchell-Jones, A.J. and McLeish, A.P. (eds) (2004) *Bat Workers' Manual* (3rd edn). JNCC, Peterborough.

Murphy, S.E., Greenaway, F. and Hill, D.A. (2012) Patterns of habitat use by female brown long-eared bats presage negative impacts of woodland conservation management. *Journal of Zoology* 28: 177–183.

Murray, K.L., Britzke, E.R. and Robbins, L.W. (2001) Variation in search-phase calls of bats. *Journal of Mammalogy* 82: 728–737.

NE (2012) Bats and onshore wind turbines (Interim guidance) (TIN051). Natural England, Sheffield. Available from: http://publications.naturalengland.org.uk/publication/35010 (accessed 18 December 2015).

NE (2013). Guidance on the capture and marking of bats under the authority of a Natural England licence. WML-G39 (10/13). Available from: http://webarchive.nationalarchives.gov.uk/20140605090108/http:/www.naturalengland.org.uk/Images/wmlg39_tcm6-35872.pdf (accessed 18 December 2015).

Palmer, E. (2013) A Study on the Population Size, Foraging Range and Roosting Ecology of Bechstein's Bats at Grafton Wood SSSI. A report to The People's Trust for Endangered Species and Worcestershire Wildlife Trust. Available from: http://www.worcswildlifetrust.co.uk/sites/default/files/files/Nature%20Reserves/Grafton%20Wood%20Bechsteins%20Report%20FINAL.pdf (accessed 20 December 2015).

Park, K.J., Masters, E. and Altringham, J.D. (1998) Social structure of three sympatric bat species (Vespertilionidae). *Journal of Zoology* 244: 379–389.

Park, K.J., Jones, G. and Ransome, R.D. (1999) Winter activity of a population of greater horseshoe bats (*Rhinolophus ferrumequinum*). *Journal of Zoology* 248: 419–427.

Parsons, K.N. and Jones, G. (2003) Dispersion and habitat use by *Myotis daubentonii* and *Myotis nattereri* during the swarming season: implications for conservation. *Animal Conservation* 6: 283–290.

Parsons, K.N., Jones, G., Davidson-Watts, I. and Greenaway, F. (2003a) Swarming of bats at underground sites in Britain – implications for conservation. *Biological Conservation* 111: 63–70.

Parsons, K.N., Jones, G. and Greenaway, F. (2003b) Swarming activity of temperate zone microchiropteran bats: effects of season, time of night and weather conditions. *Journal of Zoology* 261: 257–264.

Parsons, S. and Jones, G. (2000) Acoustic identification of 12 species of echolocating bat by discriminant function analysis and artificial neural networks. *Journal of Experimental Biology* 203: 2641–2656.

Racey, P.A. and Swift, S.M. (1985) Feeding ecology of *Pipistrellus pipistrellus* (Chiroptera: Vespertilionidae) during pregnancy and lactation. *Journal of Animal Ecology* 54: 205–215.

Racey, P.A., Swift, S.M., Rydell, J. and Brodie, L. (1998) Bats and insects over two Scottish rivers with contrasting nitrate status. *Animal Conservation* 1: 195–202.

Ransome, R.D. (1997) *The management of greater horseshoe bat feeding areas to enhance population levels*. English Nature Research Report Number 241. English Nature, Peterborough.

Ransome, R.D. and Hutson, A.M. (2000) *Action Plan for the Conservation of the Greater Horseshoe Bat in Europe* (Rhinolophus ferrumequinum). Nature and Environment, No. 109. Council of Europe Publishing, Strasbourg.

Razgour, O., Hanmer, J. and Jones, G. (2011) Using multi-scale modelling to predict habitat suitability for species of conservation concern: the grey long-eared bat as a case study. *Biological Conservation* 144: 2922–2930.

Razgour, O., Whitby, D., Dahlberg, E., Barlow, K., Hanmer, J., Haysom, K., McFarlane, H., Wicks, L., Williams, C. and Jones, G. (2013) *Conserving Grey Long-Eared Bats (Plecotus austriacus) in our Landscape: a Conservation Management Plan*. Available to download from the Bat Conservation Trust (http://www.bats.org.uk).

Reckardt, K and Kerth, G. (2007) Roost selection and roost switching of female Bechstein's bats (*Myotis bechsteinii*) as a strategy of parasite avoidance. *Oecologia* 154: 581–588.

Richardson, P. (2000) *Distribution Atlas of Bats in Britain and Ireland,* 1980–1999. Bat Conservation Trust, London.

Rivers, N.M., Butlin, R.K. and Altringham, J.D. (2005) Genetic population structure of Natterer's bats explained by mating at swarming sites and philopatry. *Molecular Ecology* 14: 4299–4312.

Rivers, N.M., Butlin, R.K. and Altringham, J.A. (2006) Autumn swarming behaviour of Natterer's bats in the UK: Population size, catchment area and dispersal. *Biological Conservation* 27: 215–226.

Robinson, R.A. and Stebbings, R.E. (1997) Home range and habitat use by the serotine bat, *Eptesicus serotinus*, in England. *Journal of Zoology* 243: 117–136.

Russ, J.M. (2012) *British Bat Calls: A Guide to Species Identification*. Pelagic Publishing, Exeter.

Russ, J.M. and Montgomery, W.I. (2002) Habitat associations of bats in Northern Ireland: implications for conservation. *Biological Conservation* 108: 49–58.

Russ, J.M., Briffa, M. and Montgomery, W.I. (2003) Seasonal patterns in activity and habitat use by bats (*Pipistrellus* spp. and *Nyctalus leisleri*) in Northern Ireland, determined using a driven transect. *Journal of Zoology* 259: 289–299.

Rydell, J., Bushby, A., Cosgrove, C.C. and Racey, P.A. (1994) Habitat use by bats along rivers in north east Scotland. *Folia Zoologica* 43: 417–424.

Schofield, H., Messenger, J., Birks, J. and Jermyn, D. (2002) *Foraging and Roosting Behaviour of Lesser Horseshoe Bats at the Ciliau, Radnor*. Report to The Vincent Wildlife Trust, Herefordshire. Available from: http://www.swild.ch/Rhinolophus/Ciliau%20Report.pdf (accessed 5 December 2015).

Scott, C. and Altringham, J. (2014) *WC1015 Developing Effective Methods for the Systematic Surveillance of Bats in Woodland Habitats in the UK*. University of Leeds, Leeds.

Scottish Government (2014) *Scottish Planning Policy*. Available from: http://www.gov.scot/Resource/0045/00453827.pdf (accessed 18 December 2015).

Shiel, C.B., Shiel, R.E. and Fairley, J.S. (1999) Seasonal changes in the foraging behaviour of Leisler's bats (*Nyctalus leisleri*) in Ireland as revealed by radio-telemetry. *Journal of Zoology* 249: 347–358.

Skalak, S.L., Sherwin, R.E. and Brigham, R.M. (2012) Sampling period, size and duration influence measures of bat species richness from acoustic surveys. *Methods in Ecology and Evolution* 3: 490–502.

Slack, G. and Tinsley, E. (2015) Linking bat surveys with meteorological data: a way to target operational wind farm mitigation. *In Practice* 87: 34–38.

Smith, P.G. and Racey, P.A. (2002) *Habitat Management for Natterer's Bat*. Mammals Trust UK, London. Available from: http://ptes.org/wp-content/uploads/2014/06/nattererbook.pdf (accessed 5 December 2015).

Smith, P.G and Racey, P.A (2005) The itinerant Natterer: physical and thermal characteristics of summer roosts for *Myotis nattereri* (Mammalia: Chiroptera). *Journal of Zoology* 266: 171–180.

Smith, P. and Racey, P. (2008) Natterer's bats prefer foraging in broad-leaved woodlands and river corridors. *Journal of Zoology* 275: 314–322.

Sowler, S. and Middleton, N. (2013) Bat passes – redundant or still useful? An alternative approach to the analysis and interpretation of large amounts of data. *In Practice* 79: 16–18.

Speakman, J.R., Webb, P.I. and Racey, P.A (1991) Effects of disturbance on the energy expenditure of hibernating bats. *Journal of Applied Ecology* 28: 1087–1104.

Speakman, J.R., Irwin, N., Tallach, N. and Stone, R. (1999) Effect of roost size on the emergence behaviour of pipistrelle bats. *Animal Behaviour* 58: 787–795.

Swift, S.M. (1997) Roosting and foraging behaviour of Natterer's bats (*Myotis nattereri*) close to the northern border of their distribution. *Journal of Zoology* 42: 375–384.

Swift, S.M. and Racey, P.A. (2002) Gleaning as a foraging strategy in Natterer's bat *Myotis nattereri*. *Behavioural Ecology and Sociobiology* 52: 408–416.

Taake, K.H. (1984) Strukturelle Unterschiede zwischen den Sommerhabitaten von Kleiner und Grosser Bartfledermaus (*Myotis mystacinus* und *M. brandtii*) in Westfalen. *Nyctalus* 2: 16–32.

Thomas, D.W. (1995) Hibernating bats are sensitive to nontactile human disturbance. *Journal of Mammalogy* 76: 940–946.

Thomas, D.W., Fenton, M.B. and Barclay, R.M.R. (1979) Social behaviour of the little brown bat, *Myotis lucifugus*. *Behavioural Ecology and Sociobiology* 6: 129–136.

Thompson, M.J.A. (1992) Roost philopatry in female pipistrelle bats *Pipistrellus pipistrellus*. *Journal of Zoology* 228: 673–679.

Underwood, A. J. (1997) *Experiments in Ecology: Their Logical Design and Interpretation Using Analysis of Variance*. Cambridge University Press, Cambridge.

Vaughan, N., Jones, G. and Harris, S. (1997) Habitat use by bats (Chiroptera) assessed by means of a broad-band acoustic method. *Journal of Applied Ecology* 34: 716–730.

Walters, C.L., Freeman, R., Collen, A., Dietz, C., Fenton, M.B., Jones, G., Obrist, M.K., Puechmaille, S.J., Sattler, T., Siemers, B.M., Parsons, S. and Jones, K.E. (2012) A continental-scale tool for acoustic identification of European bats. *Journal of Applied Ecology* 49: 1064–1074.

Waters, D., Jones, G. and Furlong, M. (1999) Foraging ecology of Leisler's bat (*Nyctalus leisleri*) at two sites in southern Britain. *Journal of Zoology* 249: 173–180.

Welsh Assembly Government (2009) *Technical Advice Note 5: Nature Conservation and Planning*. National Assembly for Wales, Cardiff. Available from: http://gov.wales/docs/desh/policy/100730tan5en.pdf (accessed 5 December 2015).

Welsh Government (2014) *Planning Policy Wales* (Edition 7). Welsh Government, Cardiff. Available from: http://gov.wales/docs/desh/publications/150924planning-policy-wales-edition-7-en.pdf (accessed 5 December 2015).

Williams, C. (2010) *Biodiversity for Low and Zero Carbon Buildings: A Technical Guide for New Build*. RIBA Publishing, London.

Williams, C.A. (2001) The winter ecology of *Rhinolophus hipposideros*, the lesser horseshoe bat. PhD thesis, Open University.

Wray, S., Wells, D., Long, E. and Mitchell-Jones, A. (2010) Valuing bats in ecological impact assessment. *In Practice* 70: 23–25.

Yau, N. (2011) *Visualize This: The Flowing Data Guide to Design, Visualization, and Statistics*. John Wiley & Sons, Indianapolis.

Zeale, M.R.K. (2011) Conservation biology of the barbastelle (*Barastella barbastellus*): applications of spatial modelling, ecology and molecular analysis of diet. PhD thesis, University of Bristol, Bristol.

Zeale, M.R.K., Davidson-Watts, I. and Jones, G. (2012) Home range use and habitat selection by barbastelle bats (*Barbastella barbastellus*): implications for conservation. *Journal of Mammalogy*, 93: 1110–1118.

Zuur, A.F., Ieno, E.N. and Smith, G.M. (2007) *Analysing Ecological Data*. Springer Science and Business Media, New York.

Zuur, A.F., Ieno, E.N. and Elphick, C.S. (2010) A protocol for data exploration to avoid common statistical problems. *Methods in Ecology and Evolution* 1: 3–14.

Appendix 1. Equipment table

Table A1.1 Equipment relevant to different survey types.

Equipment	Preliminary ecological appraisals - fieldwork	Preliminary roost assessment - structures	Presence/absence survey (structures)	Roost characterisation survey (structures)	Winter hibernation survey	Preliminary ground level roost assessment - trees	Presence/absence survey of trees using PRF inspection	Presence/absence survey of trees using dusk/dawn visits	Roost characterisation survey - trees	Bat activity survey (manual)	Bat activity survey (automated/static)	Swarming survey (acoustic)	Back-tracking survey	Trapping survey (hand net)	Trapping survey (harp trap or mist net)	Radio-tagging/radio-tracking survey
Binoculars.	✔	✔			✔	✔										
Powerful torch. Preferably non-heat-producing, e.g. LED lamp, particularly in potential hibernation situations. With filter if appropriate. More information on licensing for the use of artificial lights is provided in Section 1.2.2.	✔	✔			✔	✔	✔								✔	
Headtorch. Plus spare handy in pocket for extracting bats from traps if trapping.		✔	✔	✔	✔			✔	✔	✔	✔		✔	✔	✔	✔
Small torch.		✔			✔		✔									
Caving helmet and lamp.		✔			✔											
Extendable mirror.		✔			✔		✔									
Ladder. For safe access to a suitable working platform. Follow HSE recommendations on checking/documentation and safe use. Where safe access to a suitable working platform is not available consider alternatives such as the use of a cherry picker, MEWP or scaffold tower.		✔			✔		✔									
Compass.	✔	✔			✔	✔	✔			✔	✔		✔			✔
Tape measure or laser range finder.		✔			✔											
Clinometer.		✔			✔	✔										
Temperature/humidity logger.				✔	✔						✔	✔				
Weather station to record wind and precipitation if required.											✔	✔				
Endoscope. More information on licensing for the use of endoscopes in England is provided in Section 1.2.2.		✔			✔		✔									
Collection pots with labels and disposable gloves.		✔			✔	✔	✔									
Bat handling gloves gloves. (Different types for different-sized species.)		✔		✔										✔	✔	✔
Hand-held bat detector and recorder. Heterodyne bat detectors are not acceptable for commercial surveys.	✔	✔	✔	✔		✔	✔	✔	✔	✔			✔	✔	✔	✔
Counter.			✔	✔				✔	✔							
Hand-held radios.			✔	✔				✔	✔	✔			✔	✔	✔	✔
Night-vision scopes or infrared or thermal imaging camera.			✔	✔				✔	✔							
Automated bat detector.			✔	✔							✔	✔				
GPS.	✔					✔				✔			✔	✔	✔	✔
Tree tape (logger's tape).						✔										
Tree tags, nails and a hammer.						✔										
Rope access equipment such as harnesses, ropes, carabiners, prussic loops, strops, climbing helmet etc. (or access equipment such as cherry pickers, MEWPs or scaffold towers).							✔									

Continued: Table A1.1 Equipment relevant to different survey types.

Equipment	Preliminary ecological appraisals - fieldwork	Preliminary roost assessment - structures	Presence/absence survey (structures)	Roost characterisation survey (structures)	Winter hibernation survey	Preliminary ground level roost assessment - trees	Presence/absence survey of trees using PRF inspection	Presence/absence survey of trees using dusk/dawn visits	Roost characterisation survey - trees	Bat activity survey (manual)	Bat activity survey (automated/static)	Swarming survey (acoustic)	Back-tracking survey	Trapping survey (hand net)	Trapping survey (harp trap or mist net)	Radio-tagging/radio-tracking survey
Robust kit bag.							✔									
Hand net. More information on licensing for the use of capture is provided in Section 1.2.2.				✔					✔							
Thermometer.			✔	✔	✔			✔	✔	✔			✔	✔	✔	✔
Fine scissors to cut nets if needed.															✔	
Callipers.				✔					✔					✔	✔	✔
Bat holding bags. Drawstring to be tied firmly to prevent bat escape. Bags should be hung up rather than laid on the ground. Wash bags regularly and ensure no loose threads are present that may entangle bats inside the bag.														✔	✔	✔
Mist nets*, poles, pegs and guy lines. More information on licensing for the use of mist nets in England is provided in Section 1.2.2															✔	✔
Harp traps*, guy lines and possibly, ropes. More information on licensing for the use of harp traps in England is provided in Section 1.2.2															✔	✔
Acoustic lures*. More information on licensing for the use of acoustic lures in England is provided in Section 1.2.2															✔	
Glue. Surgical or colostomy latex glues are generally safe to use for tagging bats and are temporary.																✔
Small brush or cotton bud to apply glue.																✔
Curved scissors. To cut bat fur for tagging (unless possible to part hair).																✔
Weighing scales.																✔
Portable soldering iron and solder. To solder (and start) the contacts some types of radio transmitters. Operate on gas, ensure adequate supplies																✔
Radio transmitters**. VHF radio transmitters are small enough to fix safely to a bat without affecting its welfare to enable tracking. If several bats are being tracked simultaneously frequencies should be well spaced.																✔
Receivers (and headphones**). Scanning receivers can aid the tracking of multiple bats simultaneously.																✔
Antennae. To receive radio transmitter signals/pulses**. Antennae usually need to be tuned to appropriate bandwidth. Two types - low-range omni-directional element useful for vehicle searches of lost bats. Directional Yagi type can be three- or five-element. Five-element Yagi provide the best range and more accurate direction fixes.																✔

*See Appendix 5 for more information on mist nets, harp traps and lures.
**See Appendix 6 for more information on radio transmitters, receivers/antennae.
Note: The equipment chosen for a survey should make the survey safer, easier, more efficient and more thorough. Requirements for equipment depend on the nature of the survey and nature of the site, therefore this list should be adapted accordingly. As with all equipment, manufacturer's instructions should be adhered to and training/experience may be necessary to ensure safe and effective use.

Appendix 2. Background information on bat detectors

The three main systems for converting ultrasound produced by bats into sound that we can hear are **heterodyne**, **frequency division** and **time expansion**. In addition, **full-spectrum sampling** enables the recording of ultrasound at a high sampling rate without converting frequencies to the audible range. The last three are all 'broad-band' systems that simultaneously sample all frequencies in the bat calls, which means that all bat calls can be sampled if the sampling rate of the detector is at least double the frequency that needs to be sampled, and that recordings from these systems are suitable for sonogram analysis and bat call identification. This enables measurement of call parameters, to varying degrees of precision depending on the bat detector system used, which can help to confirm species identity. Professional surveys should only be carried out using broad-band detectors.

Heterodyne

In a simple heterodyne system, ultrasound is picked up by the microphone and mixed with a signal from a tuneable oscillator in the detector which the user can adjust, normally by turning a dial on the detector. The bandwidth varies between detectors and can affect how accurately the peak frequency of bat calls can be determined, because a narrow bandwidth makes it easier to discern differences in tonal quality (linked to peak frequency) when tuning. Conversely, a wider bandwidth may result in more bats being detected. Heterodyne bat detectors are not considered suitable for commercial surveys.

Frequency division

This is normally the cheapest of the 'broad-band' systems that simultaneously monitor the full range of frequencies contained within all bat calls. A frequency division of eight, for example, refers to counting the average time spent for eight oscillations of the electrical signal (that matches the acoustic signal). The time is measured when the voltage of the transformed sound wave equals zero.[47] This measurement of time allows a calculation of the average frequency of those eight oscillations. A single (dominant) frequency is plotted for each measurement point in time, with many more frequency points recorded in full-spectrum sampling. As a result, low-amplitude bat calls will not be recorded (unlike full-spectrum recordings) if another sound source of higher amplitude is received (e.g. background noise or interference) and harmonic frequencies cannot be recorded at the same time as a higher-amplitude dominant frequency.

Sufficient frequency information is preserved using this system to enable basic sonogram analysis; recordings can be made and analysed using software that processes the recordings to give us a visual image of the sound to represent frequency against time, but not multiple frequency content and amplitude. As zero-crossing analysis only preserves a small proportion of the detail of recordable sound, it is likely that a reasonable proportion of the bat passes received by the microphone will not be recorded when data is transformed through zero-crossing analysis. This is something to assess on a site-by-site basis and revisit depending on developments in equipment.

Time expansion

Along with full-spectrum sampling (see below), this method gives the most accurate reproduction of the bat calls. In summary, the detector digitally stores the ultrasound signal, and replays it at a slower speed. The recording retains the original signal in high resolution. When the call is replayed slowly (for example, 10 times lower in frequency), it is audible to human ears. Recently developed time-expansion units do not have recording limitations (except the size of the card), and it is now possible to listen back to time-expansion recordings while continuing to record full-spectrum data, rather than having to stop sampling to listen back to previously recorded bat calls.

Full-spectrum sampling

In addition to time expansion and frequency division systems, detectors are available that record ultrasound in 'real time' using a high-speed data acquisition card (A/D card). A microphone is connected to the A/D card which records sound at very high sample rates, thus enabling high-frequency sounds to be recorded directly. These enable the production of high-resolution sonograms as with time expansion, but also real-time continuous monitoring as with frequency division, so you get the best features of both systems. One disadvantage is that the sounds outputted by the detector are not in the audible range, so it is not usually possible to hear what you are recording in the field, although new technology means that it is possible to record in real time while listening in heterodyne, frequency division or listening back in time expansion. Some models are designed mainly for long-term unattended monitoring while others can also be used hand-held in the field and may display 'live' real-time sonograms (although note that these can be distracting, causing the surveyor to miss visible behaviours).

[47] Most frequency division bat detectors do not measure zero-crossing points as the signal at zero is not quiet but includes background noise as well as internal electronic and microphone noise. A sensitivity threshold is set above this to avoid dominant background noise masking bat calls.

Appendix 3. Hazards and risks

Table A3.1 Hazards and risks associated with bat survey work and methods to remove or reduce risk.

Hazards and risks associated with fieldwork	Procedures to remove or reduce risk	Equipment to remove or reduce risk
Lone working.	Lone working should ideally be avoided wherever possible, unless the risks can be reduced to an acceptable level using a risk assessment process. If lone working is unavoidable, a buddy system (and late working procedure if appropriate) should ensure that someone knows where each surveyor is and can raise the alarm if he or she does not return when expected. Surveyors should park so that they can drive away from a site without turning. This is useful in the dark, in case of emergency, and in case of aggression.	A mobile phone (satellite phone in remote areas), map and compass should be carried. In cases where ecologists are on the same site but working remotely a two-way radio and whistle can be useful.
Tiredness.	Limit the number of surveys carried out during the week (refer to: Working Time Regulations 1998), taking into consideration travel distances, type of survey, difficulty of terrain, etc. Book accommodation with late checkout time if working late/very early. Encourage staff to check into accommodation if tired rather than driving home.	
Bad weather.	Awareness of the weather forecast.	Clothing appropriate to the local situation.
Working in the dark.	Surveyors should familiarise themselves with the site during daylight hours.	Powerful torch (and spare torch, batteries and bulbs).
Working in confined spaces.	Confined spaces training (see Section 2.7).	Specialist equipment such as breathing apparatus, gas monitors, access tripod, winch and harnesses as appropriate to specific confined space following assessment.
Working underground where there may be sudden drops, changes in roof height, unstable rock, decaying fixtures.	Mine safety training (see Section 2.7).	Protective warm clothing, strong boots, helmet and helmet-mounted lamp. Ladders and/or ropes.
Working at height.	(Refer to Working at Height Regulations 2005). Tree climbing and aerial rescue course (see Section 2.7). Training in use of ladders or MEWPs as relevant.	Safe means of access, e.g. MEWPs, or ropes.
Working on busy roads, on railways, or on farmland with working agricultural machinery.	Highways Agency training (roads) or Personal Track Safety training (railways). If appropriate, ensure local workers know that a survey is under way.	Fluorescent or reflective jacket (appropriate to site) and other PPE as directed by client.
Working in derelict structures / construction sites / trees where there is risk of falling masonry or branches.	As appropriate, seek advice from a structural engineer on derelict buildings, gain a CSCS card for work on construction sites or for work on trees seek advice from an arborist. Ensure local workers know that a survey is under way.	Hard hat, fluorescent or reflective jacket, safety footwear.
Working near water (rivers, streams, ditches, lakes, canals, etc.).	Take care when moving around. Employ safe methods of crossing watercourses such as rivers, streams and ditches. Check flood conditions online. Work in pairs.	Life jacket (consider self-inflating type to allow for greater mobility).

Hazards and risks associated with fieldwork	Procedures to remove or reduce risk	Equipment to remove or reduce risk
Working near unfenced slurry or silage pits, ponds, grain silos and stores.	Surveyors should take due care and familiarise themselves with the site during daylight hours.	Torch or head torch.
Slips, trips and falls on rough ground.	Take care when moving around, ensure visibility is adequate. Be aware of reduced concentration when using electronic devices.	Torch or head torch.
Sunburn / sunstroke.	Awareness of the weather forecast.	Sun screen, hat, long-sleeved shirt and drinking water.
Diseases such as Weil's disease, Lyme disease, ornithosis[48] and tetanus (e.g. from rusty barbed wire).	Awareness of diseases, e.g. surveyors should carry a Weil's disease awareness medical card and be familiar with tick identification. Tetanus inoculation.	Protective clothing. Bandages or plasters over any open cuts or wounds. Ornithosis – protective dust mask and gloves.
Insect bites and stings (horseflies, ticks, etc.).	Understand the habitat preferences of different insects; be aware of insect behaviour; avoid obvious nests.	Insect repellent and/or barrier clothing (long sleeves and trousers, nets, etc.). Carry antihistamine if likely to react strongly to bites/stings.
Poisonous plants (e.g. giant hogweed).	Be able to identify these plants; don't touch them.	Wear appropriate PPE.
Bat bite and rabies (European Bat Lyssavirus).	All those who handle bats should be vaccinated (and regularly boosted) against rabies because of the risk of European Bat Lyssavirus. Care should be taken when handling to avoid bites. Information on vaccinations and what to do if bitten is available on the GOV.UK website,[49] or by calling its Centre for Infections.[50] See also the Department of Health's 'Green Book' *Immunisation Against Infectious Disease 2006* from the GOV.UK website.[51]	Appropriate gloves should be worn when handling bats (advice is available from the BCT).
Asbestos, fibreglass and dust.	Every non-residential building should have an Asbestos Register. Surveyors should ask to see it, particularly if the building being surveyed was built between 1950 and 1985. Asbestos should be avoided and a specialist asbestos consultant called if necessary.	Asbestos – disposable overalls and respirator. Fibreglass and dust – protective dust masks (conforming to BS EN149), safety glasses and overalls.
Sharp objects, such as broken glass or hypodermic syringes.	Take care when moving around, ensure visibility is adequate.	Safety work boots with protective toecaps and reinforced soles, impact-grade gloves, overalls, first aid kit.
Land that has been sprayed.	Surveyors should ask landowners or agents whether pesticides have recently been used on land being surveyed. Many pesticides have a 'harvest interval' between spraying and harvesting; surveys should not take place until after this interval.	

[48] An infectious disease that affects birds and can affect humans and other mammals.
[49] https://www.gov.uk/government/collections/rabies-risk-assessment-post-exposure-treatment-management
[50] 020 8200 4400
[51] https://www.gov.uk/government/collections/immunisation-against-infectious-disease-the-green-book

Hazards and risks associated with fieldwork	Procedures to remove or reduce risk	Equipment to remove or reduce risk
Aggressive farm animals such as guard dogs, geese, bulls and cows with calves.	Surveyors should ask landowners or agents where animals are kept, and avoid those areas if possible.	
Shooting, e.g. for predator control (often takes place at dusk).	Surveyors should ask landowners or agents when any shooting is likely to be taking place, and avoid surveying at those times. Be aware of the potential for illegal shooting.	
Verbal and physical assault.	Avoid lone working; work within sight of an accompanying surveyor; park so as to be able to leave quickly. Ask for security personnel in higher-risk areas, which could be identified through contact with the police. Withdraw as soon as practicable if risk is greater than anticipated.	Fluorescent or reflective jacket. Attack alarm.

NOTE: Unsafe work should not be carried out and ecologists should stop work if a survey becomes unsafe and consider alternative approaches to minimise risks.

Appendix 4. Protocol for bat dropping collection for DNA analysis

1. Dropping samples should be collected using clean tweezers or, if unavailable, gloves should be worn (or a sample bag turned inside out) to avoid contamination. Care should be taken to avoid breaking droppings during collection.

2. If droppings of various ages are present, those that appear most recent and most intact should be selected for analysis.

3. Where it is believed that different species are present, or droppings are present in different locations, these should be collected in separate containers and using different materials to avoid cross-contamination.

4. Although single droppings are accepted for analysis, if possible it is advisable to send at least five droppings in one sample, in case a retest is needed. However, it is also advisable for the sender to retain a few in the unlikely event of loss in transit.

5. Containers should be clean and dry, sterile if possible, but this is not essential.

6. The smallest container that will hold the sample is preferred, to avoid droppings disintegrating in transit. Ideal containers are 2.0 ml Eppendorf-type plastic tubes, or small (preferably 10 cm × 14 cm) resealable plastic bags (Ziploc or similar) are suitable. Samples can be padded with clean non-fluffy material (e.g. paper) to reduce movement in transit. Do not use glass tubes.

7. Ensure samples are labelled and packaged according to the instructions provided and that a separate note is kept by the sender of which sample numbers relate to which sample locations.

8. The sample should be dispatched to the lab as soon as possible, but if this cannot be done immediately, then it should be stored in a dry, cool place. Freezing or refrigeration is not necessary. If the sample is particularly fresh and is damp, the droppings should be air dried on a clean sheet of paper at room temperature, to help preserve the DNA and to prevent the droppings becoming squashed together in transit.

Appendix 5. Background information on mist nets, harp traps and lures

Mist nets

Specialist bat mist nets are manufactured by a range of suppliers and have smaller pockets compared to nets designed to catch birds, although this type of net can also be used. Nets come in a range of sizes, from 2m to 25m in length and 2 to 3m in height, and usually 36mm mesh. Net selection will depend on the habitat. For mist netting in closed woodlands, 6 × 2.6m nets are usually more than adequate when used in combination with an acoustic lure. Shorter nets would be more appropriate for tunnel entrances and, for more open woodlands, 9 to 18m nets can be used effectively. The height of the mist net is governed by the habitat being surveyed, and limited by pole lengths. Guy lines and pegs are also required to stabilise the net. Specialist mist nets such as canopy net systems are also available where it is necessary to work at these heights. However, the advantage of using an acoustic lure is that bats that usually occupy this habitat zone can be drawn to the traps. The main advantage of using mist nets is that the equipment is relatively lightweight and inexpensive; the trapping area is also higher than for harp traps. The main disadvantage of mist nets is that bat extraction is more difficult and thus more risky to the bat's welfare. This in turn requires greater levels of skill and training to be able to use this equipment safety and effectively. In addition nets are required to be continually monitored to limit the amount of time bats are in the net.

Harp traps

Harp traps are generally more limited in size than mist nets (usually no larger than $4m^2$). They are also more expensive and are relatively heavy items of equipment, which is an important consideration when planning the appropriate size of the team. However, their main advantage is that once captured, bats are held in relative safety and the process of collecting bats from a harp trap is less stressful for the bat and safer for the ecologist. Therefore ecologists need less training than those using mist nets. In addition harp traps do not need continuous monitoring and can be checked every 15 minutes or sometimes even less frequently, subject to licensing guidance and/or requirements, weather conditions and time of year.

Acoustic lures

Acoustic lures are devices or systems that emit recorded or synthesised social and echolocation calls of bats. Used in combination with mist nets or harp traps, acoustic lures can increase capture rates of bats significantly. Some devices are single unit and compact with built-in amplifiers and sequencers emitting synthesised calls and/or previously recorded calls of bats with either built-in or connected ultrasonic speakers. This makes them portable and easier to manage in the field and protect from the elements. Other systems include the combined use of laptop computers, high-speed sampling devices, amplifiers and ultrasonic speakers to emit recorded bat calls. The laptop-based system provides a flexible platform to alter and change calls in the field; however, the levels of equipment involved often require constant attention and exacerbate the logistical challenges. Common to all systems is that they are expensive. The use of spinning devices can increase the effectiveness of ultrasonic calls emitted by a static speaker by reflecting the highly directional ultrasonic calls in different directions, adding Doppler shift into the call and simulating a moving bat. However, the construction of these needs careful consideration to ensure that any bat that may come into contact with it cannot be injured by the mechanism.

Appendix 6. Background information on radio transmitters and receivers/antennae

Radio transmitters (tags) are the key component of a radiotelemetry system. The weight of the tag should not exceed 5% of the body weight of the bat. Lighter tags usually result in a reduction of power and lifetime of the transmitters. Depending on the configuration, the majority of bat tags generally have a life of between 7 and 25 days and at ground level a range of 1–3km when the bat is outside its roost. The range of transmitters is considerably reduced when a bat is within its roost. Transmitters can be configured in three ways by tag suppliers:

1. Transmitter antenna length can be ordered to a specified length, which should be selected depending on the size and foraging behaviour of the species or project methodology. Shorter antennae (10–15cm) reduce range but are less likely to be tangled with the antennae of other bats. These are recommended for use with smaller and close commuting species and when many bats are being tagged simultaneously. Longer antennae (15–25cm) are best used with further-ranging species and very small numbers of bats, such as when the priority is to find roosts.

2. Range to battery life ratio. Suppliers of transmitters are able to increase the power of the transmitter, which increases range at the expense of battery life. Therefore if a survey only requires tracking for a week, tags can be adjusted to reduce the battery life to 7 days, and increase the transmission power to improve the detection/location range.

3. Contact connection method – two methods are generally used for UK bat species. Reed switches are contacts within the housing on the transmitter that are held apart by the use of a magnet taped to the tag. When the magnet is removed, the tag activates, and vice versa. Reed switches make starting tags a very simple exercise in the field. However, they can be less reliable than soldered contacts, and are generally heavier. Soldered contacts are more reliable but take some skill to use in the field, require extra soldering equipment, and once connected they are harder, if not impossible, to stop. An alternative method for starting tags is the 'wire loop' method, although this is less commonly used in the UK.

At least one receiver, antenna and radio transmitter is required to undertake a radiotelemetry survey. Consideration should also be given to vehicle-mounted antennae and masts to increase the effectiveness of receiving signals at range and keeping in contact with the bat. For species with known long flying ranges, such as noctule and barbastelle, vehicle-mounted antennae are usually essential and should be anticipated as part of the survey design.

Appendix 7. Introduction to data analysis

Transformation of data

An example of transforming data is when bat passes per night are transformed into bat passes per hour to facilitate a comparison of data collected in different months with different night lengths. This is illustrated in Appendix 8, Worked Example 3.

Visualisation of data

Graphical visualisation

Graphical tools are typically used for data exploration, and to aid interpretation of the data.

A good way of comparing two or more data sets is the box plot (see Figure A7.1). The box plot visualises the median[52] and the spread of the data: the horizontal line in the box is the median, with the 25% and 75% quartiles forming a box around the median that contains half the observations. Any points outside the box are labelled outliers (outliers are retained for the analysis).The box plot in Figure A7.1 shows common and soprano pipistrelle data (1,942 records of bat passes per night) from a recent study (Mathews *et al.*, 2015). Table A7.1 gives descriptive statistics for this data set, e.g. mean, median, max, etc. These are two useful methods to summarise large data sets.

Figure A7.1 Example of a box plot.

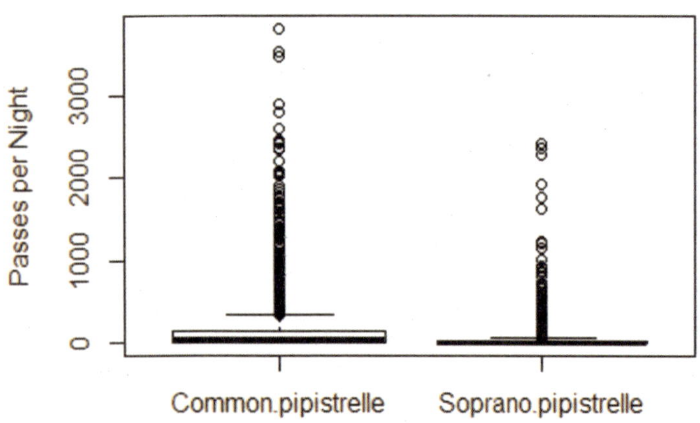

Table A7.1 Descriptive statistics for common and soprano pipistrelle passes per night.

Statistic	Common pipistrelle	Soprano pipistrelle
Number of records/nights	1,942	1,942
Mean	164.31	42.02
Median	37	5
Standard deviation	359.86	158.73
25% quartile	6	0
75% quartile	136	24
Maximum	3,815	2,426
Minimum	0	0

Box plots are one way of showing large data sets succinctly but, as shown in Figure A7.1, their usefulness may be limited where there is a large spread in data. Others methods of presenting data are the dot plot or Cleveland plot, the histogram and the density plot. Examples are given in Figure A7.2 to Figure A7.4; all show the same common pipistrelle data and all visually describe the distribution of bat passes recorded during the study.

Figure A7.2 Example of a dot plot or Cleveland plot (note that this is a one-dimensional graph with the data spread vertically to facilitate visualisation).

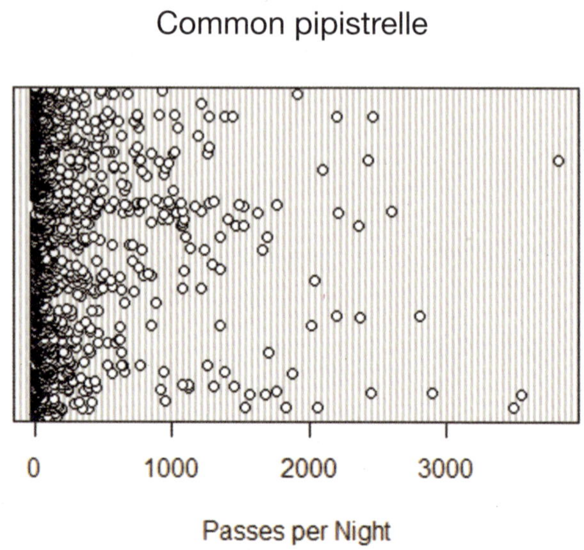

Figure A7.3 Example of a histogram (there were over 1,000 occasions when between 0 and 100 passes per night were recorded, etc.).

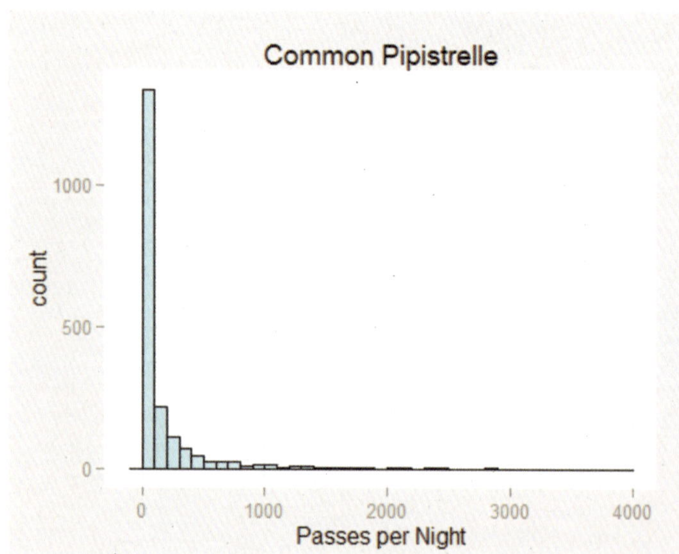

[52] The value or quantity lying at the mid-point of a frequency distribution of observed values. To find the median by hand, place the numbers in value order and find the middle number; if there are two middle numbers, average them.

Figure A7.4 Example of a density plot (similar to the histogram).

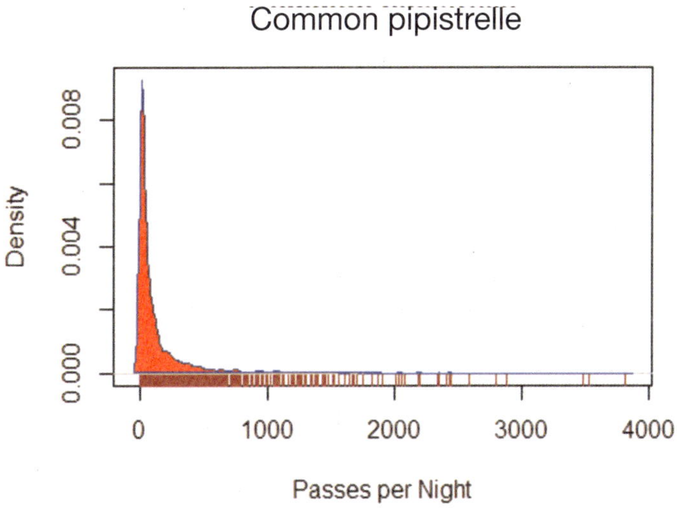

Geographical visualisation

With the rise of GPS, commonly built into bat detectors, there is more data with latitude and longitude coordinates attached; this makes maps the intuitive way to visualise the information. The examples in Figure A7.5 and Figure A7.6 present information recorded from a transect undertaken with a bat detector that records latitude and longitude and bat activity.

Figure A7.5 Geographic data is shown at the location where the bat was recorded and colour-coded according to species.

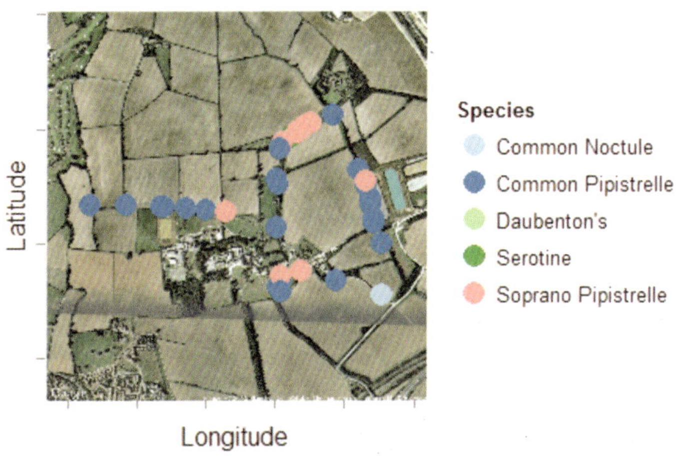

Figure A7.6 Geographic data is shown as a kernel density plot, which estimates the smoothed distribution of bat activity (Kahle and Wickham, 2013). White areas show a lower density of passes whereas red areas show a higher density of passes.

Visualisation of large data sets

The use of automatic bat detectors that operate for extended time periods, and identification software that can rapidly process the information, results in the collection of large amounts of data. Visualisation of the data is the primary way of communicating the information and its interpretation to others; it also helps in the analysis by showing the information in a readable form, something a table cannot always achieve. The difficult part is visualising the information without reducing any of the detail. The graphs in Figure A7.7 and Figure A7.8 present bat data recorded at six locations for five nights each month from May to September, as part of a wind farm proposal. The collision risk included is based on NE's Technical Information Note TIN051 (NE, 2012), which is being updated at the time of writing.

Figure A7.7 Box plot showing bat data per month recorded at six locations for five nights between May and September (log scale).

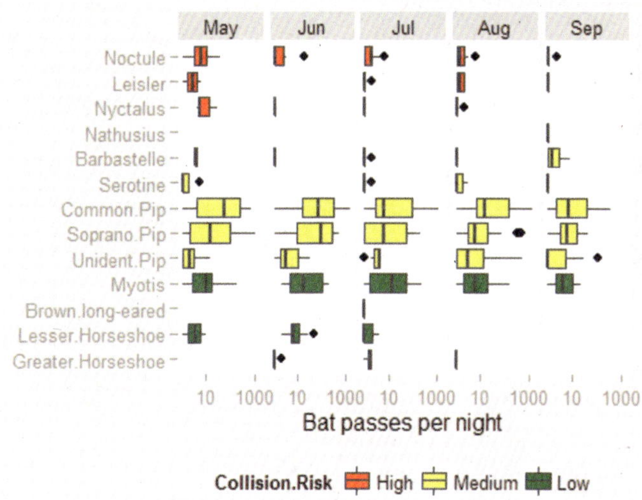

Figure A7.8 Box plot showing bat data per site recorded for five nights each month between May and September (log scale).

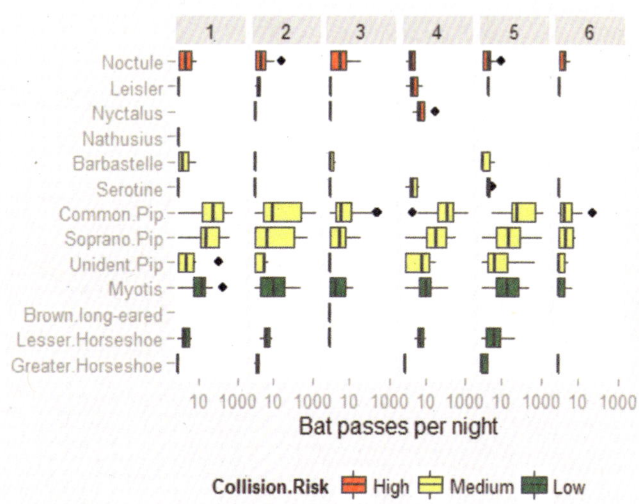

Figure A7.9 is a shaded graph that shows bats recorded for the turbine and hedge in Worked Example 2 below.

Figure A7.9 Shade plot of turbine and hedge data.

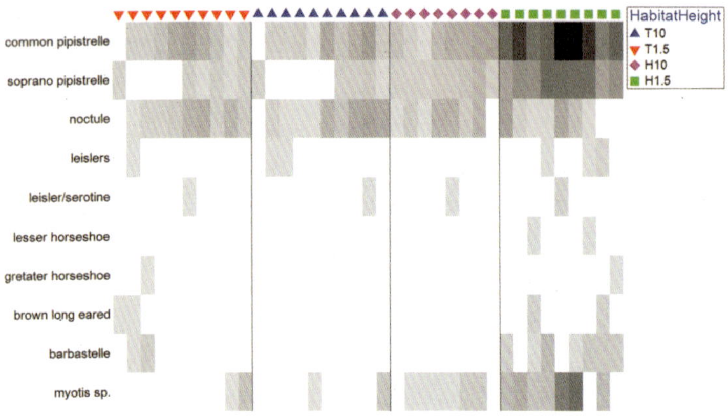

T10 = turbine at 10m, T1.5 = turbine at 1.5m, H10 = hedge at 10m, H1.5 = hedge at 1.5m. Darker shading indicates larger numbers of bats.

For a more comprehensive review of displaying information see Yau (2011).

Modelling/statistical testing

Non-parametric statistical methods are mathematical procedures for statistical hypothesis testing that make no assumptions about the distribution of the variables being assessed – **the observed data can be used as it comes**. A justification for the use of non-parametric methods is simplicity. Moreover, they leave less room for improper use and misunderstanding. Non-parametric methods are frequently suitable for processing biological data (Fowler *et al.*, 1998).

Hypothesis testing

Ecologists can use **hypothesis tests** for evidence-based assessments. The idea is that a hypothesis is formalised into a statement such as 'soprano pipistrelle activity is different at the hedge and turbine', appropriate data are collected and then statistics are used to determine whether the hypothesis is true or not. For every hypothesis there will be an associated null hypothesis and most statistical tests use the null **hypothesis** as a starting point.

For the example hypothesis 'soprano pipistrelle activity is different at the hedge and turbine', the associated null hypothesis is 'soprano pipistrelle activity is *not* different at the hedge and turbine'. What a statistical test determines is the probability that the null hypothesis is true (called the *P*-value). If the probability is low then the null hypothesis is rejected and the original hypothesis accepted.

Setting a hypothesis is a good way of not over-interpreting the data, because it defines a formal question, tests the question and provides an answer from which a defendable inference can be made.

Type I and II errors

In theory, the null hypothesis is either true or false, but only if all the individuals in a population (or a complete measure of the index) are sampled. The statistical test can only give an indication of how likely it is that the null hypothesis is true **based on the sample available**. There are two ways of making the wrong inference from the test; these two types of error, by convention, are called Type I and Type II errors, as described in Table A7.2.

Table A7.2 How Type I and Type II errors can arise in statistical testing.

Null hypothesis	Accepted	Rejected
True	Correct	Type I error
False	Type II error	Correct

In a Type I error, the null hypothesis is really true (i.e. soprano pipistrelle activity is not different at the hedge and turbine) but the statistical test has led us to believe that it is false (i.e. there are different activity levels). This type of error can be seen as a **false positive**.

In a Type II error the null hypothesis is really false (soprano pipistrelle activity is really different at the hedge and turbine) but the test has not picked up this difference. Small sample sizes will often lead to a Type II error.

P-values

The lower the probability (*P*-value) the more confidence there is that the null hypothesis can be rejected. However, unless the whole population is measured there can never be complete certainty. It is the usual convention in biology to use a critical *P*-value of 0.05. This means that the probability of the null hypothesis being true is 0.05 (5% or 1:20). In other words, it indicates that the null hypothesis is unlikely to be true.

Mann–Whitney U test and Kruskal–Wallis rank sum tests

Two frequently used tests include the **Mann–Whitney U test** and the **Kruskal–Wallis rank** sum test.

The Mann–Whitney U test is a non-parametric technique for comparing the medians of two unmatched samples. It may be used with as few as four observations in each sample. Because the values of observations are converted to their **ranks**, the test may be applied to a wide range of variables (e.g. ordinal or interval scales). The test is also distribution-free – it is suitable for data that are not normally distributed, for example the counts of bats above. Sample size can be unequal.

The Kruskal–Wallis rank sum test is a simple non-parametric test to compare the medians of three or more samples. It can be used to test any number of groups. This test *may* be used when there are only two samples, but the Mann–Whitney U test is more powerful for two samples and should be used in preference.

Appendix 8. Worked examples of statistical analysis

Worked Example 1: Turbine and hedge data

Bat surveys were undertaken at a proposed wind turbine location in SW England during the months of August and September. Four automatic bat detectors were placed for five nights in August and September at the locations shown in Figure A8.1 and Table A8.1. The survey design observes bat activity for height (1.5m and 10m) and habitat (hedge and turbine); these are all sampled equally for five nights each month.

Figure A8.1 Survey design to sample at two heights and in two habitats at a proposed wind farm site.

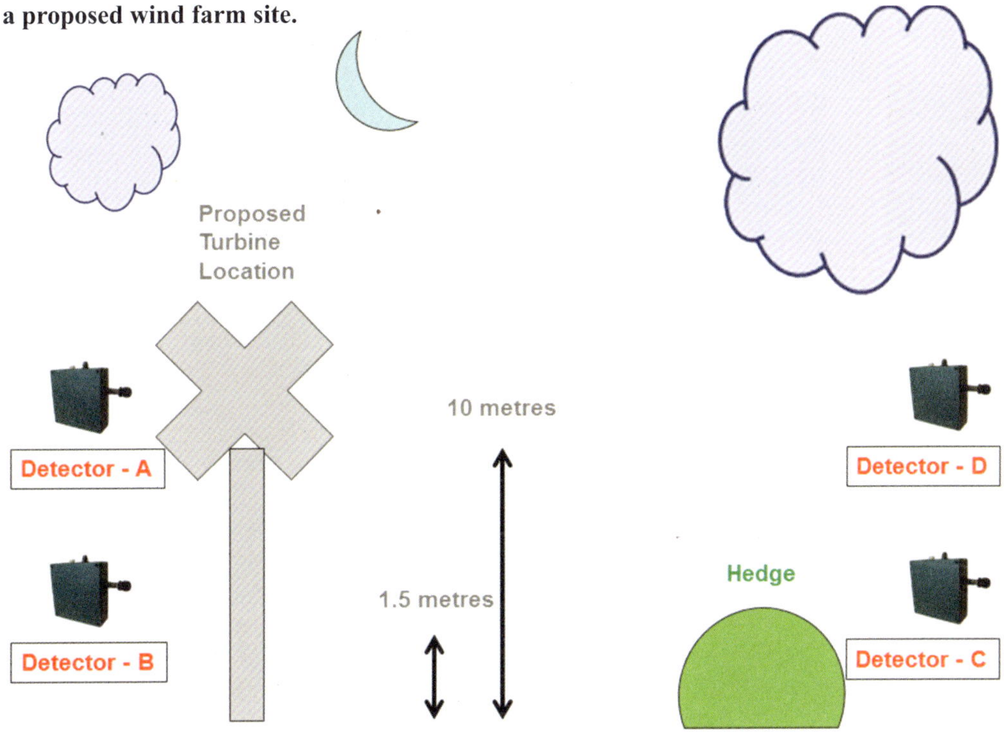

Table A8.1 Bat detector locations in relation to survey design in Figure A8.1.

Bat detector	Height	Location
Detector – A	10m	Turbine
Detector – B	1.5m	Turbine
Detector – C	1.5m	Hedge
Detector – D	10m	Hedge

The automatic detector survey measured an index of bat activity (i.e. the number of bat passes per night). The criteria for a bat pass are not important here, as long as all four locations use the same method for determining a bat pass and that method is reported.

This example describes a non-parametric approach to undertaking statistical analysis of bat survey data (i.e. the number of bat passes recorded over a set period for individual bat species).

The assumptions made are as follows:

1. The four bat detectors are considered equal in their ability to detect bats.
2. Bat species are equally likely to be detected at a given distance, e.g. loud bats such as the noctule and quiet bats such as the brown long-eared.
3. Identification of bats using sound analysis is correct.
4. The null hypothesis will be rejected when the *P*-value turns out to be less than 0.05 (5%).

The question asked is as follows:
Is there a difference between the level of bat activity at the hedge and turbine as measured by a bat pass per night index?

To illustrate the example bat passes allocated as soprano pipistrelle and noctule bat have been used.

Always start with a graph – see Figure A8.2 and Figure A8.3.

Figure A8.2 Box plot of soprano pipistrelle activity at the hedge and turbine.

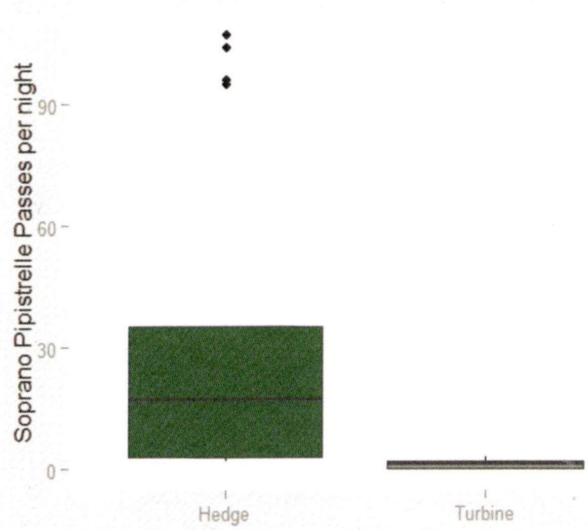

Figure A8.3 Box plot of noctule bat activity at the hedge and turbine.

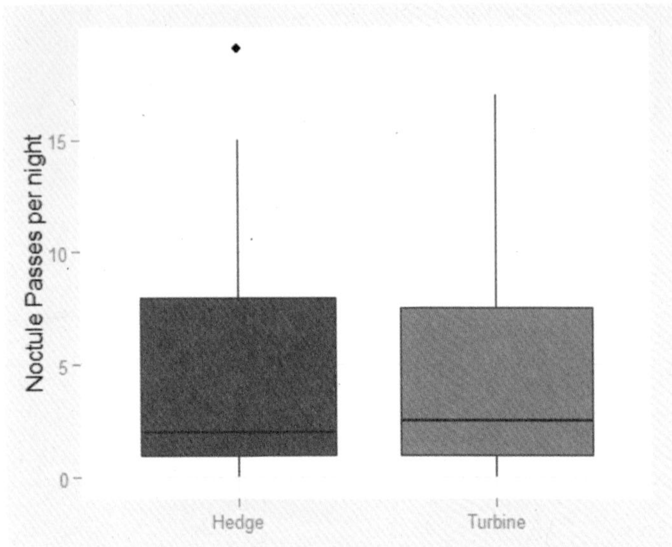

The box plots above clearly show that there is a large difference between the level of soprano pipistrelle activity at the hedge and the turbine but a much smaller difference between the level of noctule bat activity at the hedge and turbine. However, reporting of the results will be much more defensible if testing is carried out to find out whether these differences are statistically significant. The Mann–Whitney U test can be used to test these differences.

- **Null hypothesis:** hedge and turbine soprano pipistrelle (and noctule bat) activity (as measured by passes per night) come from distributions with the same median, i.e. they are not significantly different.

- **Alternative hypothesis:** hedge and turbine soprano pipistrelle (and noctule bat) activity come from distributions with a different median, i.e. they are significantly different.

When this example is put through the Mann–Whitney U test the following results are obtained.

For soprano pipistrelle, the resulting *P*-value (< 0.05) tells the ecologist to **reject** the null hypothesis, i.e. hedge and turbine soprano pipistrelle activity is significantly different.

For noctule bat, the resulting *P*-value (> 0.05) tells the ecologist they **cannot reject** the null hypothesis, i.e. hedge and turbine noctule bat activity are not significantly different.

The ecologist, using the statistically supported evidence of similar noctule activity at the turbine and hedge, could suggest that an alternative location for the turbine is investigated. Using multivariate statistical techniques (Zuur *et al.*, 2007; Legendre and Legendre, 2012) it would be possible to investigate the assemblage of all bat species observed at the four detector locations. For example, an ANOSIM test (Clarke, 1993) shows that there is a significant difference ($P < 0.001$) between the assemblage of bats at the hedge (1.5m) and the other three locations (hedge 10m, turbine 1.5m and turbine 10m).

Worked Example 2: Comparing levels of bat activity on a transect

It is possible to carry out simple quantitative analysis of bat activity data to compare the distribution of bats; for example, in different broad habitat types or in different areas within a site. This can be done using a simple chi-square test to investigate whether or not bat activity is distributed as expected from the relative sizes of the habitats or areas (Fowler *et al.*, 1998; Dytham, 2011).

The method involves assigning recorded bat activity into the different sections on the transect to be investigated, measuring the relative lengths of those sections and comparing the bat activity actually observed within each section to the activity expected if bats were randomly distributed across all of the habitats surveyed.

For example, once a transect has been planned on a site, a walkover and/or aerial photographs (e.g. from Google Earth[53]) can be used to section the transect into broad habitat categories such as:

- woodland;
- woodland edge;
- hedgerows;
- pasture.

The length of each section in each habitat is measured. The bat activity within each section can then be quantified. The expected values for bat activity are calculated based on the relative length of each habitat covered by the transect, and compared to the observed values using a chi-square test (Fowler and Cohen, 1990). In this example, illustrated in Table A8.2, the number of bat passes in each of four habitat types is shown along with the length of each habitat within a 6km transect.

Table A8.2 An example of transect survey data transformed to enable statistical analysis using a chi-square test.

Data for common pipistrelle	Woodland	Woodland edge	Hedgerow	Pasture
Transect length in habitat (km)	1	2	1.5	1.5
Observed no. of bat passes	4	21	15	2
Expected no. of bat passes	7	14	10.5	10.5

The chi-square statistic is calculated as follows:

$$\chi^2 = \Sigma \frac{(\text{Observed - Expected})^2}{\text{Expected}}$$

In this example $\chi^2 = 13.59$. A chi-square distribution table shows that bat activity is not randomly distributed between the habitat types as the result is significant ($P < 0.01$, df $= 3$). Further analysis can be completed to discover which habitats differ in terms of bat activity, or qualitative interpretation can be made from the relative levels of observed and expected activity. Table A8.2 shows that common pipistrelle activity is higher than expected in woodland edge and hedgerow habitats, and lower than expected in woodland and pasture habitats.

[53] http://www.google.co.uk/intl/en_uk/earth/index.html

Assumptions are made when completing a chi-square test which must be met before any analysis is carried out. In particular, it is assumed that the expected values for the majority of categories are > 5, and therefore the test is not suitable for species or species groups where low levels of activity are recorded.

The G-test is an alternative to the chi-square test. The two methods are interchangeable; if a chi-square test is appropriate then so too is a G-test and the assumptions in each are the same. The outcome of the G-test is a test statistic (G) which is compared with the distribution of chi-square in the same tables as the chi-square test. So why use the G-test? It is easier to calculate by hand but importantly it has been shown to be superior on theoretical grounds to the chi-square test; so the G-test should be preferred (Fowler *et al.*, 1998; Dytham, 2011).

The G-test is calculated as follows:

$$G = 2 \text{ x} \sum \text{aObserved x ln} \left(\frac{\text{Observed}}{\text{Expected}}\right)$$

where:

$\sum$ a = the sum

ln = natural logarithm

When G has been calculated as described above, the Williams' correction must also be applied. The correction factor is calculated as follows:

$$\text{Correctionfactor} = 1 + \frac{(a^2 - 1)}{6nv}$$

where:

a = the number of categories

n = the total sample size

v = the number of degrees of freedom

The adjusted G (or Gadj) is calculated as follows:

$$\text{Gadj} = \frac{G}{\text{Correctionfactor}}$$

In the example above, Gadj is 13.7544366. A chi-square distribution table shows that bat activity is not randomly distributed between the habitat types as the result is significant ($P < 0.01$, df = 3).

The analysis above shows that common pipistrelle activity is higher than expected in woodland edge and hedgerow habitats, and lower than expected in woodland and pasture habitats. Further analysis can be completed to discover which habitats differ in terms of bat activity, or qualitative interpretation can be made from the relative levels of observed and expected activity.

Worked Example 3: Nathusius' pipistrelle monthly activity

Data has been collected observing Nathusius' pipistrelle bat passes per night for each month from April to October; moon illumination was also recorded for four ranges: 0–25%, 26–50%, 51–75% and 76–100%. Table A8.3 shows the median bat passes per night for each month and moon illumination.

Table A8.3 Median bat passes per night by month and moon illumination.

Month	Moon 0–25%	Moon 26–50%	Moon 51–75%	Moon 76–100%
April	3.00	NA	2.00	2.75
May	3.30	5.0	5.70	4.50
June	4.00	6.0	12.50	5.50
July	1.00	4.0	4.00	3.00
August	3.00	1.5	2.25	2.00
September	16.00	2.0	18.00	11.00
October	38.75	0.5	14.50	5.00

NA = no data available

The question is:
Is there a difference between the level of Nathusius' pipistrelle bat activity for each month as measured by a bat pass per hour index?

To treat the months equally, the data needs to go through transformation, from bat passes per night to bat passes per hour (due to the difference in night length between the months; see Figure A8.4) by dividing monthly night bat activity by the average monthly night length (hours). This produces activity per hour for all observations, allowing the months to be compared with each other. The transformed data is presented in a box plot in Figure A8.5.

Figure A8.4 Average night-time lengths for different months in study.

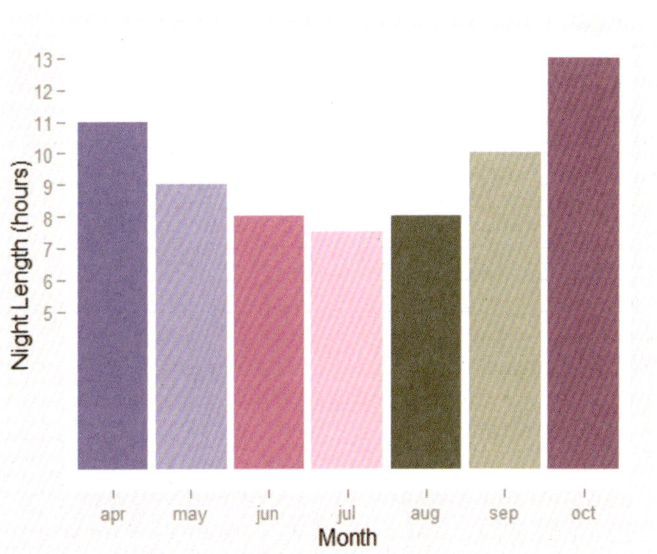

Figure A8.5 Box plot showing Nathusius' pipistrelle activity by month (passes per hour).

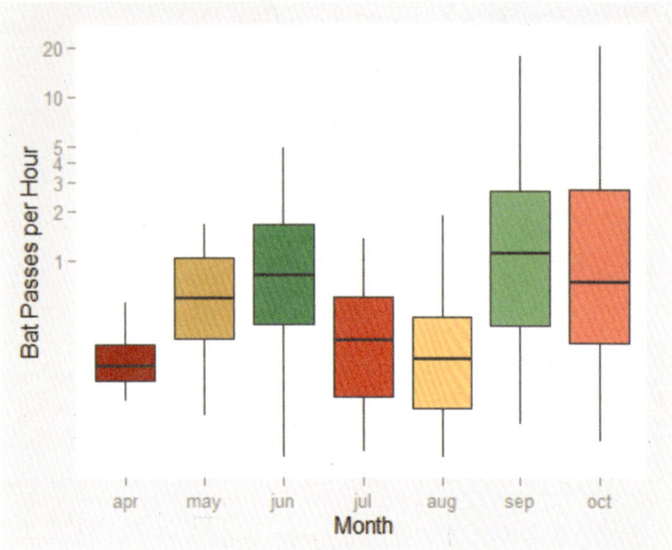

The box plots above appear to show differences in bat activity in the different months. However, reporting of the results will be much more defensible if testing is carried out to find out whether these differences are statistically significant, i.e. clearly show there is a difference. The Kruskal–Wallis rank sum test, which is a simple non-parametric test to compare the medians of three or more samples, can be used to test for differences.

- **Null hypothesis:** the Nathusius' pipistrelle passes come from distributions with the same median, i.e. they are not significantly different between months.
- **Alternative hypothesis:** the Nathusius' pipistrelle passes come from distributions with a different median, i.e. they are significantly different between months.

When this example is put through the Kruskal–Wallis rank sum test the following result is obtained.

The resulting *P*-value of 0.0018 (i.e. < 0.05) tells the ecological consultant to reject the null hypothesis, i.e. Nathusius' pipistrelle passes are significantly different between months. This test tells us that there is a significant difference between the months, but not which month or months. Further testing would be required to investigate which pairs of months are significantly different using further Kruskal–Wallis rank sum testing (Field *et al.*, 2012). When carried out, the further test shows that there is a significant difference in activity ($P < 0.05$) only between the months of August and September.

The objective was to look for evidence of Nathusius' pipistrelle peaks in activity during a given month or group of months. September is believed to be a key time for migration and Nathusius' pipistrelle activity was found to be significantly higher in September than August. However, statistical testing showed that activity of this species was not higher in September than in any of the other months. There may be other factors involved, for example temperature, wind speed or rain, or more data may be needed either from further fieldwork or by combining data from other similar studies; of course it may mean there really isn't a difference. It could also be possible that the sample size is too small, leading to a Type II error.

Worked Example 4: Nathusius' pipistrelle activity and moon illumination

Using the data from Worked Example 3 we investigate the levels of activity of Nathusius' pipistrelle with different moon illuminations.

The question:
Is there a difference between the level of Nathusius' pipistrelle bat activity for each category of moon illumination as measured by a bat pass per hour index?

A box plot of the data is given in Figure A8.6.

Figure A8.6 Box plot showing Nathusius' pipistrelle activity by moon illumination.

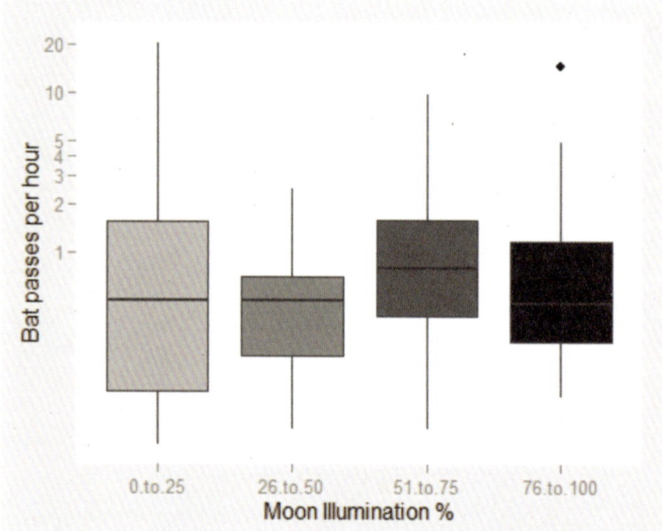

The box plot above appears to show differences in bat activity according to different moon illuminations. However, reporting of the results will be more defensible if we test whether these differences are statistically significant – is there a clear difference? The Kruskal–Wallis rank sum test can be used to test the differences.

- **Null hypothesis:** the Nathusius' pipistrelle passes come from distributions with the same median, i.e. they are not significantly different between moon illuminations.
- **Alternative hypothesis:** the Nathusius' pipistrelle passes come from distributions with a different median, i.e. they are significantly different between moon illuminations.

When this example is put through the Kruskal–Wallis rank sum test the following result is obtained.

The resulting *P*-value of 0.339 (i.e. > 0.05) tells the ecological consultant to not reject the null hypothesis, i.e. Nathusius' pipistrelle passes are not significantly different between moon illuminations.

Note: If the box plot showed a trend in activity with increasing moon illumination, we may want to make a hypothesis to explain this trend and then test it. The Jonckheere–Terpstra statistic can be used to test for an ordered pattern (increasing or decreasing) in the medians of the four illumination levels. It is similar to the Kruskal–Wallis test, but incorporates information about whether the order of the groups is meaningful (Field *et al.*, 2012). This test may be particularly useful for, for example, post-construction monitoring purposes – detecting a year on year increase/decrease/no change of activity.

Index

*Note: where there are several page numbers the main page number(s) are shown in **bold** type; an 'n' suffix to a page number refers to a footnote or a note to a table.*